Forbes Winslow

Light: Its influence on life and health

Forbes Winslow

Light: Its influence on life and health

ISBN/EAN: 9783337269272

Printed in Europe, USA, Canada, Australia, Japan

Cover: Foto ©berggeist007 / pixelio.de

More available books at **www.hansebooks.com**

LIGHT:

ITS INFLUENCE

ON LIFE AND HEALTH.

BY

FORBES WINSLOW, M.D.,

D.C.L. Oxon. (Hon.)

&c. &c.

NEW YORK:
MOORHEAD, SIMPSON & BOND, PUBLISHERS.

1868

TO

SIR WILLIAM FERGUSSON,

Fellow and Member of the Council of the Royal College
Surgeon and Professor of Surgery at King's College H
traordinary to Her Majesty the Queen, and Surge
His late Royal Highness the Prince Consort;
of Human Anatomy and Surgery, Royal
Surgeons, England, and Examiner o
gery at the University of Londo

THIS WORK IS DEDIC

BY THE AUTHOR,

AS A TRIFLING MARK OF ADMIRATION A

FOR

HIS DISTINGUISHED TALENTS AND BRILLIA

AND

IN PLEASING REMEMBRANCE

OF AN UNINTERRUPTED FRIENDSHIP OF
YEARS.

PREFACE.

THE object of this work is to demonstrate the inestimable value of light as an hygienic agent, and to analytically examine its physiological influence in the development of vital phenomena as manifested in the animal and vegetable kingdoms.

How impossible it is to exaggerate the amount of blessing which flows from the action of the sun, that

> "*Great source of day, for ever pouring wide
> From world to world the vital ocean round,*"

not only on organic, but inorganic matter!

Deprived of its life-generating and health-sustaining power, the whole of animated nature would be a sterile blank, and man, the highest order of intelligent beings, become blighted in mind and degenerated in body.

In the composition of this volume I have been under the

necessity of levying contributions on the writings of established physicists.

I lay no claim to original experimental research, but desire only to be considered as a wanderer on the vast shore of science, picking up here rare shells, there valuable pebbles, and then classifying them, with a view of illustrating a branch of inquiry of deep philosophical interest.

It is impossible to contemplate the facts scattered through the following pages without being solemnly impressed with a sense of man's deep obligation to God, the Source of all true LIGHT, for thus beneficently surrounding our planet by an influence so pre-eminently conducive to the health and happiness of the human race.*

CAVENDISH SQUARE, LONDON,
April, 1867.

---

* A portion of the chapter on the "Lunar Ray" appeared many years ago in a paper of my own, on "Lucid Intervals," published in the *Psychological Journal*.

TABLE OF CONTENTS.

PART I.

THE SOLAR BEAM.

	PAGE.
Origin of light,	1
Healthful and morbid effects of light upon the animal and vegetable kingdom,	2
No vitality or healthful structure without light,	3
"Solar-air-bath" of the ancients,	3
Bodily deformities, crime, and mental disease produced by darkness,	4
Health of those who live in the country as compared with the inhabitants of manufacturing towns,	4
The absence of light on miners and colliers,	5
Diseases caused by the absence of light,	6
Unhealthy condition of those who labor by night and sleep by day.	7
Deep mining at Geelong,	7
Development of children retarded by working in mines,	8
Stunted forms observed among the miners of Chimay,	9
Effects of dampness and deficient nourishment viewed in connection with want of light,	9
Influence of light upon animals,	10
Arrest of animal development,	11
Blindness of fish caused by want of light,	12
Discoloration of different races of men caused by solar heat,	14

	PAGE
The brown skins of the natives of Lapland and Greenland,	14
White skins of negroes dwelling in the forests,	15
Black color contrasted with that of red or olive negroes,	15
Black, white, and spotted negroes,	16
Effect of food on the hair,	16
Complexions of intemperate persons,	16
Red Indians turned white,	17
Influence of the sun in tropical climates,	18
Albinos of the torrid zone,	18
Fish of great oceanic depths,	18
Effect of the absence of light on the women of the seraglio,	19
Statures of different races of men,	19
Mental condition of the Arabs,	22
Inhabitants of high latitudes possess superior muscular force,	22
Northern countries more favorable than southern to longevity,	23
Influence of climate upon physical and mental organization,	23
Effect of deep-sea water on vegetable and animal life,	24
The depth of the ocean,	25
Depth at which solar light penetrates,	27
Temperature of the ocean,	28
Different colors of the sea,	29
Causes of the varied tints of the sea,	30
Distribution of animal life in the sea,	30
Classification of marine animals,	31
Microscopic shells found in deep sea deposit,	32
Animal and vegetable life incompatible with certain altitudes and depths,	33
Amount of pressure at certain oceanic depths,	33
Tendency of plants to follow light,	35
Plants and animals nurtured in darkness altered in color,	37

	PAGE
Animal and vegetable substances near the surface of the sea brilliantly colored,	39
Brilliant colors of animals and plants in the tropics,	40
Action of light necessary for development of vegetable colors,	42
Physiological influence of light on the colors of plants,	44
A definite quantity of heat required for development of plants,	44
Insect life,	47
Chemical influence of light on certain drugs,	49
Effect of light passing through different colored media on the germination of plants,	52
Metals discovered in the solar beam by the spectroscope, "Fraunhofer's lines,"	54
Diseases caused by exposure to intense solar heat,	58

PART II.

THE LUNAR RAY.

Opinions of the ancients as to the influence of the moon,	59
Different parts of the body under planetary influence,	64
The moon's power in causing tidal action,	67
Belief of Spaniards and West India negroes that the sick always die at ebb-tide,	67
Principal authorities in support of the theory of planetary influence,	68
Drs. Orton and Balfour's view of the subject,	71
Periodicity as exhibited in the phenomena of life,	76
Historical analysis of lunar influence,	78
Dr. Mead's views of the influence of the moon upon the bodily health stated at length,	78

	PAGE
Dr. Pitcairne's case of lunar influence,	89
Mr. Cockburn's singular case,	90
Hemorrhage from the finger said to come on at the full of the moon,	90
Baglivi's singular case,	91
Remarkable case of lunar influence occurring to Mr. Ainsworth,	91
Van Helmont on the influence of the moon on asthma,	91
Historical analysis of Dr. Balfour's treatise on the influence of the moon,	92
Mr. Francis Day's observations on the influence of the moon at Madras,	96
Dr. Orton on the effect of the moon on the diseases of tropical climates, and on the weather,	100
Dr. Kennedy's views on the same subject,	101
Diemerbroeck on the influence of the moon on the plague,	101
Dr. Lardner's opinion of the influence of the moon on the epidemic fever of Italy in 1693,	103
Dr. James Johnson on the effect of the moon on the diseases of the liver,	104
Dr. Scott on planetary influence in modifying diseases of tropical climates,	104
Mr. Hutton on the effects of the moon at Prince of Wales's Island,	105
Dr. Moseley on the effect of the varied phases of the moon on hemorrhages of the lungs,	106
Remarkable case illustrative of the fact,	107
Nicholas Fontana on the influence of the moon on the fever that occurred at Barrackpore in 1777,	108
Dr. Millingen on planetary influence,	109
Opinions of the Druids of Gaul on the influence of the moon in vegetable productions,	110
The mistletoe gathered with a golden knife when the moon was six days old,	110

	PAGE
The verbena of the ancients supposed to be under planetary influence,	111
Its medicinal and sacerdotal virtues,	111
Effect of different phases of the moon on the fructification of plants,	111
Instructions for planting seeds according to the position of the moon,	113
The lunar rays calorific,	115
Effect of moonlight on the thermometer,	115
Sudden death and coma caused by the moon's light,	116
Effect of full moon in causing diseases of the brain,	116
A stroke of the moon,	117
Opinion of the Arabs as to the morbid influence of the moon,	117
On the *modus operandi* of the lunar light,	118
Distance of the earth from the sun,	118
Velocity of planetary light as compared with that of sound,	119
Humboldt's remarks on lunar distances,	120
Comparison of the light of sun and moon,	120
Michel and Euler on the same subject,	121
Arago's observations on the polariscope,	121
Uses of the polariscope,	122
Sir David Brewster on polarized light,	123
State of the barometer at different phases of the moon,	125
The influence of the moon on the weather,	128

PART III.

ON THE ALLEGED INFLUENCE OF THE MOON ON THE INSANE.

"Lucid intervals," origin of the term,	129
Legal definition of the word "lunatic,"	129

	PAGE
Meaning of the phrase "lucid interval,"	130
M. D'Aguesseau's analysis of the term,	131
Dramatic and poetic references to the influence of the moon on the mind and passions,	132
The phenomenon of periodicity viewed in relation to insanity,	136
Illustrations on the subject from the writings of Pinel,	136
Periodical attacks of drunken madness, or dipsomania,	137
The periodicity explained independently of lunar influence	139
Pinel's explanation of the phenomenon,	139
Bouillon and Sauvage's cases of lunar influence,	140
Celsus on the influence of the moon,	140
Dr. Haslam's opinion on the same subject,	141
Dr. Woodward on the influence of the moon on the insane,	143
Effects of sol-lunar influence on maniacal paroxysms,	145
Effect of the eclipse of the sun in 1806 on the lunatics in Bethlehem Hospital,	146
Daquin on the influence of the moon on the insane,	147
Guislain's opinion on the same subject,	149
Effect of meteorological phenomena on the insane,	150
Sleeplessness and agitation noticed among the insane at the full of the moon,	150
Effect of light and darkness on the illusions and hallucinations of the insane,	151
Influence of different colors on the brain and mind,	153

PART IV.

HYGIENE OF LIGHT.

Effect of the absence of light on the blood, heart, brain, and muscles,	154
Sir David Brewster on the blessings of light,	155

	PAGE
On the construction of buildings with the view to the admission of the maximum degree of light,	156
On the unhealthily-housed populations of New York and Bethnal-green,	156
The effect of the absence of light on the poor living in courts, narrow streets, &c.	157
Sir David Brewster's suggestion for remedying this evil,	157
On the physical condition of the working classes of Edinburgh,	159
Sir David Brewster on the value of light as a hygienic agent to the poor,	159
Scorbutic affections caused by the absence of light,	161
On the exclusion of light from young infants and children	162
Rules for the treatment of children with reference to their being exposed to the influence of light,	162
Dr. Andrew Winter's opinion on the subject,	163
Humboldt on the robust health of the Mexicans, Peruvians, Indians, &c., caused by their being freely exposed to light,	164
Dr. Bryson on the unhealthy condition of seamen confined in dark holds of ships, etc.,	165
Judicious hygienic arrangements made for the patients of the hospital of St. John, at Brussels,	167
Importance of light to persons suffering from disease,	167
Florence Nightingale on the construction of public hospitals,	168
The effect of darkness on disease as observed in the hospital at St. Petersburg,	170
Singular case of the same kind related by Baron Dupuytren,	171
Dr. Hammond on the importance of light to the sick,	171
The virulence of cholera increased by the absence of light,	172
Injury caused by glaring light and sudden transition from darkness to light,	173
On disorders of the vision caused by prolonged exposure to intense light,	173

	PAGE
The effect of snow and ice upon the eyes of sheep.	178
Ditto on the Greek soldiers, as related by Xenophon,	179
The effect of the glare of brilliantly-lighted rooms on the organs of vision,	179
Injuries resulting to the eyes from artificial light,	179
Medical instructions for the regulation of the different colors of light,	180
On the probable effect of the vapor of iron found in the sunlight by the spectrum on the health of the body,	181
On the introduction of metallic poison into the body through the skin,	182
APPENDIX,	185

ON
THE INFLUENCE OF LIGHT.

Part I.—The Solar Beam.

"God said, Let there be light: and there was light."* These were the sublime words of the Almighty when, by an act of Sovereign power, He willed into existence " two great lights, the greater light to rule the day, the lesser light to rule the night."

The object of the following essay is to consider in a philosophical, and I hope a reverential spirit, the physiological and pathological—the healthful and morbid—effects upon the animal and veget-

---

* ויהי אור ויהי אור—"Be light! and light was."---Gen. i. 3.

able kingdom of that principle which, at the fiat of the wise and beneficent Creator, radiated in all its original glory from the Heavenly luminary, and when seen by Him was emphatically pronounced to be "good."

This subject has two aspects: of the life-giving and benign effects of light I purpose fully to speak. It is also my intention to consider the baneful influences of the solar beam and lunar ray, upon the vegetable creation, vital force, physical and mental development.

When speaking of the unalloyed happiness in store for man, in a purer, higher, nobler, and beatified state of existence, an inspired prophet places prominently in advance among the great blessings that will be associated with our exemption from the pernicious operation of those physical agents which in this world are known (under certain conditions) to be obnoxious to our well-being, the promise, that "the sun shall not smite thee by day nor the moon by night."*

After so solemn a declaration, need I hesitate in approaching the consideration of not only the *sanitary* but the *morbid* influence of the sun upon

---

* Psalm cxxi. 6.

the human race; and need I offer any apology for analytically investigating the *ab*normal as well as the *nor*mal effects of lunar light on the *flora* of creation, the material and mental organization of man?

As it would be irrelevant for me to enter into the discussion of the varied theories of light that have been propounded, I proceed at once in the first place to establish how essential the solar beam is to the preservation of the bodily and mental health, promotion of longevity, beauty of the physical form, serenity and integrity of the mind.*

There can be no persistent vitality nor healthfully developed bodily structure without light. If it were possible for a human being to be placed during the natural term of his existence, in a posi-

---

* From the earliest periods in the history of medicine, solar heat was considered to prolong human life. "Old men," says Hippocrates, "are double their age in winter, and younger in summer." In order to obtain the full advantages of the light and heat of the sun, the ancients had terraces built on the tops of their houses called *solaria*, where they took what was termed their "solar-air-bath." Speaking of his uncle, Pliny observes, "*Post cibum æstate si quid otii, jacebat in sole.*" As the sun rose, disease, according to the views of the ancients, declined, "*Levato sole levatur morbus*," was a recognized medical axiom in former times.

tion of perfect darkness, the physical tissues and mental faculties would undergo serious modifications and degenerations.

Where light is not permitted to permeate, there are found, in the highest state of manifestation, bodily deformities, intellectual deterioration, crime, disease, early and often sudden death. A material, as well as a *moral* and *mental*, etiolation or blanching occurs when the vital stimulus of light is withdrawn.

A vast body of evidence conclusively establishes the inestimable value of this agent to the health of both body and mind. Compare the bright, ruddy, happy faces and buoyant spirits of those who reside in the country and work in the open fields, and upon whom the sun is generally shining, with the pale phlegmatic faces, emaciated, stunted forms and nervous depression of those whose vocation in life deprives them of the health-giving and beneficial influence of light. In order properly to test the question, place side by side the robust, sunny, rosy-cheeked and happy lass or lad of the village with the pallid, sickly, melancholy girl or boy residing in one of our manufacturing towns, who are pursuing their vocation in a room or hermetically sealed building, where darkness has obtained an undue supremacy. Notice the difference between the two clssses of human beings! It may be enunciated as an indisputable fact, that all who live and pursue their calling in situations where the minimum degree of light is permitted to penetrate, suffer seriously in bodily and mental health.

These pathological phenomena are principally observed among those confined in dark mines and collieries, badly-constructed houses containing but few windows, holds of ships, factories, and in persons incarcerated in dungeons and buried for a considerable length of time in prisons, as well as the denizens of narrow streets, crowded alleys, confined courts, garrets, or cellars, where the blessed light of the sun has a difficulty in penetrating.*

The total exclusion of the sun's beams from the

---

* "I never shall forget," says Dr. Hammond, "the appearance presented by the sick of a regiment I inspected about a year since in Western Virginia. They were crowded into a small room, from which the light was shut out by blinds of india-rubber cloth. Pale and exsanguined ghost-like looking forms, they seemed to be scarcely mortal. Convalescence was almost impossible; and doubtless many of them died, who, had they been subjected to the operation of the simplest laws of nature, would have recovered."—"Treatise on Hygiene," Philadelphia, 1863, p. 209.

body induces the severer types of *chlorosis* (green sickness) and other anæmic conditions depending upon an impoverished and disordered state of the blood. Under these circumstances, the face assumes a death-like paleness, the membranes of the eyes become bloodless, and the skin shrunken and turned into a white greasy waxy color.

Associated with these symptoms there are emaciation, muscular debility and degeneration, œdematous conditions, dropsical effusion, softening of the bones ("mollities ossium"), general nervous excitability, morbid irritability of the heart, loss of appetite, tendency to syncope and hæmorrhages, consumption, physical deformity, stunted growth, mental impairment and imbecility, coupled with premature old age. The offspring of those so unhappily trained are often deformed, weak, and puny, and are disposed to scrofulous affections.

The wretchedly unhealthy condition of that portion of the working population of this country who live and work in darkness was satisfactorily proved at the Parliamentary inquiry which took place on this subject many years ago.*

---

* The light in the pent-up dwellings of large towns, inhabited by the poor, is polarized light, it being always only a reflected

Similar consequences are observed in all those who labor by night and sleep by day; such as bakers, police-constables, compositors connected with the daily press, whose occupation necessitates their employment during the greater part of the night.

The data collected by foreign writers fully substantiate this truth. I refer particularly to the investigations made and facts recorded in reference to the health of the miners of Belgium, Hungary, and France, as published by Drs. Ozanam, Hoffinger, Hallé, and M. Dupectiaux.* According to the observations of these distinguished authori-

---

solar beam. The effect of the light and atmosphere contained in such dwellings on the mind as well as the body is as previously stated.

* I extract the following from the *Geelong Advertiser*. It must be taken *quantum valeat*.—" It is a curious fact, connected with deep mining, that from the hours of twelve at night till three in the morning the disturbing influence in the bowels of the earth obtains increased activity. At that time it is observed by miners that water falls from places where none is observable during the day. The volume in the wheel is perceptibly increased, the atmosphere is charged with gases which often prevent the lights from burning, and small particles of earth and rock are observed to fall from the tops of the drives. Whether this phenomena is to be attributed to the diurnal motion of the earth or other causes, it is worthy of the attention of the curious."

ties, the morbid changes on body and mind which are consequent upon the exclusion of light are singularly significant and remarkable.

In numerous instances the general health becomes vitiated, and the function of nutrition seriously interfered with. Not only are specific diseases generated, but physical as well as mental development arrested. In many young girls and boys who are at an early age consigned to a life of darkness in mines, or in other places, puberty is either never attained or is greatly retarded; in fact, the mind as well as body becomes impaired in its growth, as the effect of light being excluded from the human organism.

No doubt the damaged health of those who are constantly working in dark mines and collieries is not altogether attributable to want of light, although observing and practical men have assigned this as one of the principal reasons for the physical and mental ailments to which this section of the laboring classes is liable.

Fourcault affirms that where life is prolonged perhaps to the average term, the evil effects of the want of light are seen in the stunted forms and general deterioration of the human race. It appears that the inhabitants of the arrondissement of

Chimay, in Belgium, three thousand in number, are engaged partly as coal-miners and partly as field-laborers. The latter are robust, and readily supply their proper number of recruits to the army; while among the miners it is in most years impossible to find a man who is not ineligible from bodily deformity or arrest of physical development.*

In association with the want of light should be considered the deleterious effects of dampness, deficient nourishment, impure air, over work, anxiety of mind, the inhalation of noxious and poisonous gases, fine dust, &c., upon the general and specific health of those exposed to the injurious operation of these physical agents.

With these prefatory remarks I open the subject of light in its physiological and morbid relations. I would, however, premise that it is not my object to discuss the special astronomical phenomena of the solar ray; neither do I propose to weigh the relative value of the corpuscula, undulatory, or other theories of light. I desire to restrict my remarks to the direct influence of this agent on vital manifestations, and to refer to some of the

---

* Causes Générales des Maladies Chroniques. Paris. 1844.

more remarkable facts illustrative of the effects of the exclusion of light on the bodily and mental health, and on animal and vegetable life.

In the first place, I have to point out the influence of light on the vital and physical organization of different animals. Dr. Edwards instituted a number of interesting experiments with a view of ascertaining the effect of light, independently of heat, upon the structure and growth of animals, particularly between the interval of conception, fecundation, and adult age.*

This physiologist placed some frog's spawn in water. The vessel containing it was made impervious to light by being covered with dark paper. A similar experiment was tried in another vessel which was transparent. The two vessels were exposed to the same degree of temperature, but the transparent one was placed under the influence of the sun's rays. The eggs were developed in succession; of those in the dark none did well. Unequivocal indications were observed of the transformation of the embryo in a few of the eggs. Dr. Edwards' experiments were confined to the ba-

---

* "On the Influence of Physical Agents on Life." By W. F. Edwards, M. D., F. R. S. 1832.

trachian family. The absence of light did not necessarily interfere with the development of these reptiles. Two tadpoles, out of twelve confined in a tin box, pierced with small holes for the change of water, and placed at the depth of several feet in the Seine, underwent the regular change of organic form.*

It appears that the transformation of two of the animals was retarded by the absence of light. For the purpose of fully settling this question, tadpoles were confined in two large vessels, each containing ten litres of water, both capable of admitting light; one of glass, but with a partition close to the water, to prevent atmospheric respiration; the other open, in order to allow the animals an opportunity of rising to the surface, and inflating the lungs. Those who were deprived of fresh air were later in transforming themselves than the others, but this delay was so short that the interference with the respiration appeared to be too slight to

---

* "I have," says Dr. Hammond, "often repeated Dr. Edwards' experiments, and always with analogous results. On one occasion I prevented for 125 days the development of a tadpole, by confining it in a vessel to which the rays of light had no access. On placing it in a receptacle open to light, the transformation was at once commenced, and was effected in fifteen days."

produce any effect on vital development. It was clear that the exclusion of the light had a serious influence on the transformation of the tadpoles plunged in water.

Sir Humphrey Davy refers in one of his most charming works to the arrest of development that takes place in the *Proteus Anguinus* consequent upon its exclusion from light.*

In the grotto of the Madalena, he says, at Adelsburg, in Illyria, many hundred feet below the surface, are seen creatures like slender fish, moving in the mud below the water. These are the protei: the animal is of a fleshy whiteness, and transparent in its natural state, but when exposed to light its skin gradually becomes of a darker color, and at last gains an olive tint. Being abundantly furnished with teeth, it is inferred that the animal is one of prey, yet in its confined state it has never been known to eat, and it has been kept alive for many years by occasionally changing the water in which it was placed. They have also been found at a place thirty miles distant from the cavern, thrown up by water from some subterraneous cavity. In dry seasons the protei are very

---

* "Salmonia; or Consolations in Travel."

seldom seen in the lake of the Madalena, but after great rains they are often abundant. Their natural residence is an extensive subterraneous lake, from which in great floods they are sometimes forced through the crevices of the rocks into the place where they are found.

These singular creatures Sir H. Davy affirms have no organs of vision, but in their place are two small dots which occupy the position of eyes. It has not been ascertained that they have any power of perception. The entire absence of color, and the imperfect development of their organs, in at least their intermediate condition, between those of a reptile and a fish, seem to be the result of the absence of light.

Their exclusion from the solar beam is well known to produce organic alterations in the visual organs of animals, such as atrophy of the optic nerve, or those portions of the brain (the *corpora quadrigemina*) more immediately associated with the sight. It is supposed that the blindness observed among fish found in the dark caves of the Tyrol and Kentucky arises from the arrest in the development of the eyes as the result of a constant deprivation of light.

In elaborating this subject I have to consider

the effect of solar light and heat in discoloring the skin of different races of men by producing minute injections of the capillary vessels on the surface of the body and developing what physiologists term black *pigment* cells. These peculiarities, however, are not confined to shades of color, degrees of physical development, stature, and power, but extend to the animal passions as well as mental energy. "Is not," asks Mr. Hunt, " the short-lived beauty of the Oriental women to be attributed to the influence of that sun

"'Shining on, shining on, by no shadow made tender,'

which is known to give all the grandeur to the vegetable world of the East?"

Black, brown, and copper-colored skins are observed among those who reside in tropical climates in proportion to the intensity of the solar light, and the degree to which the body is exposed to its influence.

This discoloration of the skin is not, however, perceived among those who live in temperate and cold regions.

As we approach near to the pole the skin assumes a browner cast.

This is evidenced among the Laplanders, Esquimaux, and Greenlanders. In the arctic regions there cannot be said to exist, in the right acceptation of the term, any night. Here a constant light prevails, if not from the sun yet reflected from the snow and ice or emitted by the aurora borealis. According to Captain Ross the natives of these climes suffer from the effects of intense light, and are subject to inflammation of the eyes and blindness caused by cataract and paralysis of the optic nerve. These diseases are not confined to the natives.

Even among the African negroes the skin is observed to lose its intense black color, particularly in those whose occupation enables them to live for many months of the year in the shades of the forest where the sun's rays, owing to the thickness of the foliage and the size of the trees are unable to penetrate.

It is said that the black color of the negro race is less permanent in its character than the red or olive tints. The children of olive and copper-colored parents, according to the authority of Dr. Mason Good, exhibit the parental hue from the moment of birth, but in those of blacks it is occasionally six, eight, or ten months before the dark pigment is fully secreted. It frequently happens that the black pigment-cells are not matured even among negroes. Hence are observed in this race of men children with white skins or bodies interspersed with only interrupted lines of black patches. These are called "spotted negroes." The black pigment has been known during attacks of severe illness to be entirely absorbed, and a white pigment cell developed in its place, thus affording a remarkable illustration of a *black person being suddenly bleached into a white man!* In the "Transactions of the Royal Society" is an account of a woman whose left shoulder, arm, and hand were as black as an African negro's, whilst all the rest of the skin was perfectly white. This anomalous condition was said to have arisen from her mother having during pregnancy put her foot upon a lobster!

No doubt the coarseness of the wool of animals, and hair of human beings, as well as the amount of color developed in their skin, is, to a degree, caused by the nature of their food. Oils and spirits are well known to disorder the liver, and those who indulge to excess in these articles of diet have generally an excess of bile circulating in the blood. Intemperate persons addicted to vinous potations have a sallow and olive-hued complexion. The dark and dingy color of the pigmy people, who live in high northern latitudes, arises principally from the fish and oils of a rancid and often offensive character upon which they mainly subsist. This kind of diet is believed not only to affect the color of the skin, but to cause a diminution in the stature of this race, in consequence of their food being difficult of assimilation and defective in nutrition.

Dr. Mason Good affirms that swine or other animals fed on madder-root, or *gallium-verum*, ("yellow-ladies'-bedstraw,") have the bones tinged of a deep red or yellow color. M. Huber, the celebrated naturalist of Lausanne, was able, by the quality of the food given, to convert what is commonly, but improperly, called a neuter, into a queen bee.

Trinocq, when speaking of the effects of light on the color of the body, affirms that among the Red Indians the skin that is not exposed to the influence of the sun's rays is quite white; in other words, the face, neck, and hands are dark, whilst the limbs which are covered by clothing are of an entirely different hue.

The same phenomena are observed in all tropical climates. The white color of animals inhabiting the polar regions is attributable to the absence of intense sunlight.

The changes referred to are clearly traceable to the *direct* influence of the sun, and are not exclusively caused by a high degree of temperature. Persons who live in hot and tropical climates, and whose occupation exposes them to great heat apart from the solar beam coming into immediate contact with the body, do not suffer from any marked discoloration of the skin.*

Among the negroes of the torrid zone, Albinos are seen ; but these become scarce near the equator. It is a commonly observed fact that among fish, those parts of the body not exposed to the free action of light become of a pale hue, and are sometimes quite white. This fact is noticed among a certain species of flat-fish. The *axolotus pisciformis, syren lacertina*, and the *triton sittatus*, (fish that live at great sea depths, and are consequently excluded from the light,) have transparent bodies.

Certain animals, whose natural hue is white, if bred and brought up in darkness, become completely altered in texture and color. The cockroach, in its normal state, is intensely black. If this insect be taken at an early stage of its existence and carefully reared in darkness, instead of

---

* Gardeners who are occupied for the greater part of the day in hot-houses, exposed to intense heat, are not generally tanned.

assuming an inky hue when it arrives at full growth, it becomes nearly white.

The larvæ of most insects that burrow in the cavities of the earth, of plants, or of animals, are white from the same cause. When confined under glasses that admit the influence of solar light, they exchange their whiteness for a brownish hue. It is said that Asiatic and African women confined to the walls of their seraglios and secluded from the sun, are as white as Europeans.

It is of a deep philosophical interest to consider in relation to the present subject, not only the effect of climate, meterological changes, combined with the influence of varied degrees of light upon the general health, great nervous centres, and muscular development, but the modifications which the physical force undergoes from the same causes as far as the growth, stature, and longevity of the inhabitants of various countries are concerned.

Investigations have been made for the purpose of ascertaining the effect of various kinds of climate upon the stature of different nations. It has been satisfactorily established that persons placed under the same solar influences have not the same stature; and, reciprocally, the same stature is found among people placed relatively to the same influence under very different conditions. The subjoined facts in illustration of this subject are quoted from M. Sanson-Alphonse's thesis on the influence of light and the development of health.

In the southern hemisphere the Papoes of Vaigion, at 1° from the equator, the Vankoros at 12°, the Bushmen mountaineers at 30°, (an Ethiopic race,) who receive the rays of the sun with great intensity of light and heat, have a stature of 5 ft. 1 in., very little above that of people who dwell in the northern hemisphere, the Kamschatdales, the Tartars, and the Esquimaux, (a Mongol race,) in latitude 60° to 70° north, where only the oblique rays of the sun reach them, people of whom the stature is 4 ft. 3 in. to 4 ft. 7 in.

High stature is met with equally in very different latitudes; thus, in the southern hemisphere we find 5 ft. 10 in. to be the mean height of the Caribs, (an American race,) who inhabit regions situated between 1° and 8°, the natives of the "Iles Marquises" at 10°, the inhabitants of the Navigator's Islands at 14°. People whose countries are very near each other differ considerably in stature. Thus the Patagonians, of whom the mean height is 5 ft. 10 in., are only separated by the narrow straits of Magellan from the inhabitants of Terra del Fuego, of whom the stature is only 4 ft. 3 in.; thus the people of Sweden and Finland join the Laplanders; and the New Hebrides, so near the Navigator's Islands, whose population have a very high stature, are inhabited by a short and badly-made race.

People of the same stature dwell either in maritime regions or on the continents up to a level sufficiently elevated above the sea. Temperature has certainly more influence upon stature than light.

The inhabitants of very cold regions, whether situated near the poles or on the tops of mountains covered with snow, are small.

Increase of stoutness is observed among people who inhabit regions far from the equator, and especially those in which there is much moisture. It is there especially that the marks of a lymphatic temperament are apparent. Under the same zones, but on more elevated plains, are seen men who combine the greatest bodily development with the maximum of muscular power. They are of a sanguine temperament.

In ascending further above the level of the sea and up to the regions of snow, the men are small, but thick-set.

As we approach the equator and elevated regions, the inhabitants are withered, with the muscular system very moderately developed, but endowed relatively with considerable energy. Such are especially the Arabs. Their constitutions present the character of a nervo-bilious temperament.

Although the free life of savages may be favorable to the regular development of their forms, they do not acquire under the influence of climate any remarkable muscular power.

The natives of Timor, New Holland, and Tasmania, are very inferior to the English sailors in trials of strength (Péron).

The natives of America show the same inferiority of physical force. The negro race, who bear in the highest degree the impress of the effects of light, always appear, although individually not the most active, the best suited for continued labor under the light and heat of the sun. Within the limits of climate inhabited by the negro race they appear to enjoy a very strong constitution. The same observation applies to the Arabs.

The inhabitants of high latitudes consume much more food than those who dwell in intertropical regions, and have more corporeal development. The superior muscular force with which they are endowed is partly explained by this great consumption of food, by their habits of labor, and the consequent increase in the volume of their muscles.

Growth is more rapid in equatorial regions, and puberty is reached earlier; but the age for marriage ceases sooner. There are more examples of longevity among the people of the north than among those of the torrid zone and more southern countries; but some particular facts would lead me to believe that these people would also attain a very advanced age.*

It is not my intention to consider the ethnological view of this subject. That the influence of solar heat materially modifies the national character cannot for a moment be doubted. Like plants and animals, man, the highest order of intelligence, obeys the laws regulating the world's zonal arrangements. According to climate, degrees of temperature, etc., the human race becomes passive, savage, doltish, or intelligent. These affect his physiognomy, dialect, and habits of life, as well as his mode of subsistence. When speaking of in-

---

* "De l'Influence de la Lumière sur le Développement et la Santé." Par M. Sanson-Alphonse. (Agrégé de la Faculté de Médecine de Paris.) Paris, 1852.

fluences of climate as well as local positions upon the physical and mental organization, a recent writer observes, after glancing at man's geographical position: " Here, active, intelligent, and progressive; there, sluggish, dull, and stationary; here, enjoying the highest amenities of civilization; there, grovelling in a condition little elevated above the brutes by which they are surrounded. And not merely do they differ in intellectual qualities, but in physical organization, in mien and stature, in form of head and expression of face, in color of skin, in strength and endurance, and, in fine, in all those purely bodily qualities by which one species of animal is distinguished from another."*

I have now to refer to the influence of deep sea-water on vegetable and animal life ("flora" and "fauna"), so far as its stratum illustrates the effect of the *exclusion of light* (under these circumstances) in modifying vital force, and in interfering with or arresting physical development.

It will, however, be necessary to consider, preliminarily, in connection with this subject, two important and interesting questions: 1st. What

---

* "Advanced State of Physical Geography." By D. Page. London. 1864.

according to well-ascertained data, is the depth of the sea in various portions of the globe? 2d. To what extent solar light is known to penetrate the ocean?

1st. It has been found, owing to the irregularity in the form of its basin and a variety of other causes, impossible to ascertain with any degree of scientific exactness the *mean* depth of the ocean. In consequence of the character of the tidal waves and certain undulatory movements which occur in deep waters, four miles has been said by some eminent physicists to be the maximum extent to which the plummet has descended.*

---

* " The *mean elevation* of all the land—continents and islands, mountains and plains—has been estimated by Humboldt at somewhat less than 1000 feet; and the *mean depth* of the ocean has been calculated by Laplace, from tidal waves and kindred phenomena, to be at least 21,000 feet, or about four English miles. We know, however, that a very large proportion of the ocean is comparatively shallow, and not a tithe of this depth; and therefore, to make up the mean, some other portions must be proportionally deeper, and to the extent it may be of eight or ten miles. Indeed, soundings (no doubt open to question) have been made in the South Atlantic, both by British and American navigators, varying from 27,000 to 46,000 feet; and soundings, perfectly reliable, have been taken in the North Atlantic, off the bank of Newfoundland, to the depth of 25,000 feet; while from calculations on the velocity of tidal waves, which are found to pro-

Reliable authorities, however, affirm that oceanic soundings have been taken at the depth of five, eight, and even to the extent of ten miles. There are numerous instances where twenty, thirty, forty, or even fifty thousand feet of line have been payed out without giving distinct evidence of its having reached the bottom of the sea. I state this on the authority of Sir J. Herschel. The average depth of the Pacific (in consequence of its being abundantly bestrewn with islands), is said to be less than that of the Atlantic Ocean.

In 27° 26′ S. latitude and 170° 29′ W. longitude, Sir James Rosse found the depth of the sea to be 14,450 feet. About 450 miles west of the Cape of Good Hope, it was 16,062 feet, or 323 feet more than the height of Mont Blanc; and 900 miles from St. Helena, a line of 27,000 feet did not reach the bottom of the sea, a depth equal to the height of some of the most elevated peaks of the Himalaya. It is, however, supposed that there are many parts of the ocean still deeper. Speaking in general terms, the depth of the At-

---

ceed according to the depth of the channel, it has been estimated that the extreme depths of the same ocean are about 50,000 feet, or more than nine miles." (Advanced Text-Book of Physical Geography. By David Page, F. R. S. E.)

lantic averages from three to four miles. The "Telegraphic Plateau," stretching from Cape Clear to Cape Race, a distance of 1640 miles, is only about 11,000 or 12,000 feet in depth; the greater depths (from 4 to 6 miles) have been determined in the Indian and Southern Oceans. The Arctic is of moderate depth, and characterized by great irregularity and diversity. The greatest ascertained depth in the Mediterranean is about 13,000 feet; in the Red Sea, 6300 feet; Baltic, 840 feet; Caribbean Sea, 14,000; and in the Gulf of Mexico, about 8000 feet.

2d. The depth at which solar light is believed to permeate the ocean is calculated to be about 700 feet.* The distance to which the eye can penetrate the sea depends principally upon its transparency.†

In experimentalizing as to the depth of the

---

* In connection with this subject it is important to notice the fact referred to by Sir J. Herschel, viz., that the light which penetrates to a great depth in the sea, is very different in its photochemical qualities from the solar light of the surface. Again, in calculating the degree to which the solar beam is seen to pierce the sea, it will be necessary to bear in mind the effect of various media through which light passes on the refraction of its rays.

† David Page.

ocean, it has been suggested that independently of considering the transparency of the water, we should test its phosphorescence; the face of the sky, whether clear or cloudy; the state of the sea, whether rough or smooth; the condition of the weather, whether calm or windy.

The heat of the water should also be tried, at various depths and hours of night and day, in order to ascertain not only the maximum temperature and average depth of the warmest stratum in the day, but the difference in its temperature and position by day and by night.* These observations will afford the data, also, for computing the amount of solar heat that penetrates the bosom of the sea, as well as the amount that is radiated thence again. They will reveal to us knowledge concerning its actinometry in other aspects. We shall learn how absorption by, as well as radiation

---

* Accurate calculations have been made as to the temperature of the ocean. The results obtained clearly establish that the lowest degrees of temperature are obtained on the surface of the water. About ten feet below the surface the thermometer rises several degrees. Ninety degrees is said by Mr. Agassiz (son of Professor Agassiz) to be the highest temperature he has known the ocean to attain. At very great depths of the ocean a uniform temperature of about $39\frac{1}{2}$ degrees has been found. (Dr. Page.)

from, the under strata is affected by a rough sea, as when the waves are leaping and tossing their white caps; and how by its glossy surface, as when the winds are hushed and the sea smooth.

Mrs. Somerville asserts, that in some parts of the Arctic Sea, shells are distinctly seen as low down as 80 fathoms; and among the West Indian Islands, at the same depth of water, the bed of the sea is as clear as if seen in air. At this great depth shells, corals, and seaweeds of every hue display to the eye of the observer the beautiful prismatic tints of the rainbow.

In connection with this subject it is of interest to consider the varied color of the sea in different parts of the globe, for upon this principally depends the depth at which the sight can reach. Assuming as a well-established fact that the waters of the ocean derive their color from animalcules of the infusorial kind, vegetable substances, and minute particles of matter, it may be affirmed as facts recorded by all writers on Physical Geography, that the sea is white in the Gulf of Guinea, black around the Maldives, vermilion off California, (caused by the red color of the infusoria it contains,) and green in the Persian Gulf.

In the Arctic Sea the color of the water undergoes rapid transitions from utramarine to olive green; from purity to opacity. The green color is said to be produced by myriads of minute insects which devour one another and are a prey to the whale. The varied tints of the ocean in its shallow parts depend to a degree upon the sea bed over which it passes. A bottom of chalk or white sand produces an apple green color; over yellow sand, a dark green; brown or black, over dark ground; and grey over brown mud.*

Physicists have instituted a series of interesting experiments with a view of ascertaining by carefully executed deep-sea soundings, the maximum depth at which various species of animals, (marine mollusca and those belonging to the invertebrate class,) as well as plants, are capable of sustaining life.

With this object Professor Forbes was engaged for a period of eighteen months in the Ægean Sea, and on the coast of Asia Minor, in sea bottom explorations by means of the dredge. He has demonstrated by actual observation that below the depth of 35 fathoms, the number of animals diminishes as we descend, until at the depth of about

---

* Physical Geography.

200 fathoms the number of testacea was found to be only eight, and a zero in the distribution of animal life was indicated at probably about 300 fathoms. Green fuci were not found below 55 fathoms, and millepoza not deeper than 105 fathoms. Similar phenomena are observed in some portions of the British Channel, viz., the south coast of Cornwall.*

Professor Bailey of West Point, an eminent American geographer, naturalist, and microsco-

---

* The following is the classification of marine animals according to the zones of depth at which they live:

1st. Littoral, found between high and low-water marks, a species capable of living in air as well as water, such as crustaceous animals, crabs, testacea, shell-less animals which close themselves up or seal themselves hermetically on the rocks and remain dormant during the recess of the water. Among these are found the petallas mytilli and littorinas purpuras, and among the zoöphytes the common sea anemone.

2nd. The circumlittoral zone, existing at 15 fathoms.

3rd. The median zone, from 15 to 50 fathoms.

4th. The infra-median ; and

5th. The abyssal zone ; the former from 50 to 100 fathoms, the latter from thence to the lowest depths at which life is possible.

Sir John Herschel says, when referring to this subject, that each of these zones is characterized by species which belong to no other, and each passes into the other by the intermixture of species common to several.

pist, says in a letter to a brother philosopher, "The bottom of the ocean at the depth of more than two miles I hardly hoped ever to have a chance of examining; but thanks to Brooke's contrivance, we have it (the deep-sea deposit) clean and free from grease, so that it can at once be put under the microscope. I was delighted to find that all these deep soundings are filled with microscopic shells; not a particle of sand or gravel exists in them. They are chiefly made up of perfect little calcareous shells ("*foraminifera*"), and contain also a number of silicious shells ("*diatomaceæ*"). It is not probable that these animals live at the depths where these shells are found, but I rather think that they inhabit the waters near the surface, and when they die their shells settle at the bottom." Lieut. Brooke, of the North Pacific Exploring Expedition, procured specimens in the coral sea from the bottom at the depth of 2150 fathoms. These soundings yielded representatives of most of the great groups of microscopic organisms. Professor Ehrenberg, of Berlin, found the deep-sea specimens to consist principally of calcareous *carapaces*, filled with soft pulp or fleshy matter. In Dr. Wallack's "Notes on the Presence of Animal

Life in the vast depths of the Sea," reference is made to the experiments of Capt. Sir Leopold McClintock during his survey in H. M. steamer the *Bulldog*, of the telegraphic route *viâ* Greenland. It is said this navigator brought up from the bottom of the ocean living star-fish adhering to the deep-sea line. Is this, it is asked, conclusive proof of the existence of animal life in the depths of the ocean? It is supposed that these living star-fish took hold of the line near the surface of the water, and were not actually brought up from the bottom of the sea.

The preceding facts conclusively establish the intimate relation existing between each solar beam, life and organization as far as the sea is concerned. Speaking in general terms, it may be affirmed that animal and vegetable vitality is incompatible with an altitude of a few thousand feet above the sea-level on land, and a few thousand feet beneath it in the waters. Such is the line of demarkation between organic and inorganic matter, life and sterility!

In connection with this subject it is essential to consider the amount of pressure which exists at great oceanic depths, for upon this often depends the existence of marine life. In the Arctic Sea the specific gravity of the water is lessened on account of the greater proportion of fresh fluid produced by the melting of the ice, the pressure at the depth of a mile and a quarter being 2809 pounds on a square inch of surface. Capt. Scoresby confirms this fact. He says in his "Arctic Voyages" that the wood of a boat suddenly dragged to a great depth by a whale, was found, when drawn up, so saturated with water forced into its pores, that it sank in water like a stone for a year afterward. Even sea-water is reduced in bulk from 20 to 19 solid inches at the depth of 20 miles. The compression that a whale can endure is wonderful. Many species of fish are capable of sustaining great pressure, as well as sudden changes of pressure. Divers in the pearl-fisheries exert great muscular strength, but man cannot bear the increased pressure at great depths, because his lungs are full of air, nor can he endure the diminution of it at great altitudes above the earth.*

"According to experiments, water at the depth of 1000 feet is compressed one 340th of its bulk ; and at this ratio the pressure at the depth of one mile would be equivalent to 160 atmos-

---

* Physical Geography. By Mary Somerville.

pheres, or 2320lbs. on the square inch; while at the depth of 4000 fathoms, or about 4½ miles, it would amount to 750 atmospheres! It is owing to this enormous pressure that closed bottles sunk to great depths have their corks always forced in; and that pieces of oakwood carried down to similar depths, have their fibres and pores so compressed as to be afterward incapable of floating on the surface."*

I proceeded in the next place to notice the direct as well as indirect influence of light on the vegetable kingdom, particularly in reference to the modifications effected in the coloring principle of plants, as a consequence of their exclusion from its operation; but before doing so, I would briefly refer to some interesting facts illustrative of the tendency exhibited by certain plants to follow light, from an apparently instinctive consciousnes of its being necessary, if not to their existence, at least for their well-being. This phenomenon is beautifully shown by the following facts: In the spring a potato was left behind in a cellar where some tools had been kept during the

---

* Physical Geography. By David Page, F. G. S. p. 121. London. 1864.

winter, and which had only a small aperture at the upper part of one of its sides. The potato, which lay in the opposite corner, shot out a runner which first ran twenty feet along the ground, then crept up along the wall, and so through the opening by which light was admitted.*

The *Chrysanthemum Peruvianum* turns continually toward the sun, and is said to cover itself with dewy clouds which cool and refresh the flower during the most violent heat of the day.

Professor Henslow states that some compositæ as " *Hypochæris radicata*" and " *Apirgia autumnalis*" are seen in meadows where they abound, *inclining their flowers toward the quarter of the heavens in which the sun is shining*. Vaucher says the same phenomena are observed in the *Narcissusses* and in certain species of *Melampyrum*. Similar phenomena have been observed in other plants by Priestley, Sennebier, Theodore de Saussure, Dumas, and Boussingault.

If a leafy shoot of any plant is bent down

---

* "Natural History." By Jesse.

As a general rule all plants bend toward the sun. Mr. Hunt, however, found that light transmitted through *red* fluid media had the effect of repelling plants from the solar beam. This fact, he says, is not susceptible of explanation.

without injury so as to reverse the usual position of the faces of the leaves, the latter will twist upon their petioles and *turn their upper surfaces to the light.* Shoots of slender-stemmed, quick-growing plants, such as jasmine, ampelopsis, &c., when trained into a dark recess, turn their growing point backward on the older part of the stem, also twisting round to face the light. The influence of lateral light is distinctly seen in the plants grown in rooms in front of the windows, while the drawn-up condition of closely packed trees is an equally evident result of preponderating top light.*

The curious phenomena previously referred to regarding the alteration in the color of plants caused by the absence of the solar beam are well known to all intelligent gardeners and scientific horticulturists. Plants, as well as animals, nurtured and grown in perfect darkness, never acquire their natural color. The former become white instead of green.†

This fact is observed in the *etiolation* or blanch-

---

* "Botany, Structural and Physiological." By Arthur Henfrey, F. R. S. 1857.

† " *Audi alteram partem.*" The following facts appear in a

ing, as it is termed, of certain kinds of vegetables, such as celery, seakale, endive, &c. Their leaves deprived of the sun's rays do not attain their normal growth or form, neither is the natural odor of such plants fully developed. Professor Robinson, descending into a coal-mine, accidentally met with a plant growing luxuriantly. Its form and qualities were new to him. The sod on which it grew was removed, potted and carefully attended

---

slight degree to contradict the general law just enunciated. "The *absence of light* exercises a very great influence over the power possessed by food in increasing the size of animals. Whatever arouses and excites the attention of the animal, and makes it restless, increases the natural waste of the different parts of the system, and diminishes the tendency of food to enlarge the body. To the rearers of poultry the rapidity with which fowls fatten when kept in the dark is well known; and direct experiment on other animals, whether by keeping them in the dark or by the cruel practice of sowing up their eyelids, as is adopted in India, have led to similar results. Absence of light, from whatever cause produced, seems to exercise a soothing and quieting influence on all animals, increasing their disposition to take rest, making less food necessary, and causing them to store up a greater portion of what they eat, in the form of fat and muscle." From a Paper on the "Scientific Principles involved in the Feeding and Fattening of Stock," read by Ed. W. Davey, M. B. M. R. I., at the Roy. Dub. Soc., April 14, 1859.

to in his garden.* The etiolated plant languished and died ; but the roots speedily threw out vigorous shoots which, from the form of the leaves and their peculiar odor, he readily recognized as tansey. He repeated similar experiments upon other plants, viz., lovage, carvi, and mint, with analogous results.

The *data* obtained from the study of physical geography conclusively establish that animal and vegetable substances near the surface of the sea are, in consequence of their free exposure to light, brilliantly colored, and that they gradually lose the brightness of their hues as they descend into deep sea-water, until the animals of the lowest zone are found nearly colorless. "Here," says Mr. Hunt, "an everlasting darkness prevails in the region of silence and eternal death."

The rich color of seaweed found near and on the surface of the water is owing to its free exposure to light. The same may be said of the lovely tints of some of the zoöphytes (*actinæ*) living in shallow waters, particularly the *sea-anemone*, which, under the influence of the bright

---

* "Encyclopædia Britannica."

sun, is seen whilst adhering to the rocks, to expand itself like the blossoms of a flower.

On the surface of the globe the influence of solar rays is shown in a marked manner. The animals and plants of tropical climates grow with a richness of color, far exceeding in beauty and brilliancy those of temperate zones. The latter are less exuberant in their growth, and are of a darker hue. In the arctic regions they are found nearly colorless. The changes in vital development which are observed as we approximate toward the pole are believed to be governed more by the *isothermes* than by the parallels of latitude; in other words, by the mean amount of heat diffused.*

---

* "The strength of evidence appears to be in favor of considering Light, Heat, and Actinism as three distinct principles or powers, active in regulating the great phenomena of Nature. In the sunbeam these powers are balanced against each other, and thus are determined those differences of climate which are not influenced by the physical conformation of the earth's surface.—Hunt.

It has been proved by well-conducted observations, that with variations of latitude there are variations in the relations of these three principles.

In the temperate regions of the earth the Actinic power is active; as we advance to the Tropics, where Heat increases, and

"Though vegetation may greatly differ," says Dr. Draper, "in its luxuriance in different climates of the globe, the manner of action of the light is always the same. Nothing is gained under the brilliancy of the tropical skies beyond a shortening of the time required for the accomplishment of a given amount of work. No substances are there decomposed, even in the organism of plants, which could not equally well be decomposed by the feebler light of more temperate climates, only in these it would demand more time. The oils and other substances, almost or quite free from oxygen, which abound in the

---

"The sun shines for ever unchangeably bright,"

the chemical power is weak. The photographic picture which could be taken in London in a second or two, could not be obtained within the Tropics in less than a quarter of an hour. It often happens, indeed, that prolonged exposure under a blazing sun is insufficient to produce any chemical change. Everything appears to favor the view that the distribution of plants and animals on the surface of the earth is regulated by the balance of physical forces in the sunbeam.

In the seasons we detect the same influences at work. Actinism or chemical power is greatest in the Spring; as the bright Light of Summer advances, the power of the Solar rays to produce any chemical change is diminished; and as we advance to Autumn, the peculiar Heat-rays come more evidently into action.—Mr. Hunt, in the "Popular Science Review." July, 1862.

plants of the torrid zone, are not exceptions to, but illustrations of, the doctrine here set forth."*

What is the *modus operandi* of solar light in altering if not altogether destroying the *chlorophyll*, or, to use Dr. Hope's term, the *chromogen* or coloring principle of plants? Do these changes depend, as Sir John Herschel suggests, on a "chromatic analysis by which the two distinct elements of color are separated by destroying the one and leaving the other outstanding"?

Chlorophyll is a substance similar to wax in its nature, and is contained in the deep cells or *mesophyllum* of leaves. It depends upon the action of light for its elaboration, and is intimately associated with the phenomena of active vegetable life, has a granular form, and is soluble in alcohol.†　As light decreases in the autumn the *chlorophyll* in many cases diminishes. The alteration is caused by the plant losing a portion of its carbon.

---

* "Human Physiology, Statical and Dynamical." J. W. Draper, M. D. 1858. p. 462.

† "Chromogen" is said to consist of two separate principles, one of which forms a red compound with acids, and the other yellow with alkalies. Dr. Hope attributes the green color produced by the latter to the mixture of the yellow matter with the blue infusion.

Thus may be explained the changes that take place in the evergreen and deciduous leaves, producing what is commonly known by the autumnal tints.*

The late Professor Lindley made many interesting experiments with a view of illustrating the physiological influence of light on the color of plants.† He established that the flowers of the

---

* Balfour.

† Every plant, from the time of its germination till the close of its organic activity, requires for its full development a definite quantity of heat. On this subject Schleiden has made some inquiries, and has taken as his subject the case of barley. He says "in Egypt, on the banks of the Nile, barley is sown at the end of November, and harvested at the end of February; the period of vegetation, therefore, amounts to about 90 days, and the mean temperature of this season is 69° 48'. In Tuqueres, near to Cumbal, under the equator, the time of harvest the middle of November, the mean temperature of this vegetating season of 168 days is 50° 12'. At Santa Fé de Bogotá they number 122 days between seed-time and harvest, with a mean temperature of 47° 24'. If, now, the number of days is multiplied by the figures of the mean temperature, we obtain 6282 for Egypt, 8433 36-60 for Tuqueres, for Santa Fé, 6489 48-60 therefore, as nearly the same number as the uncertainty in the estimate of the days, the accurate mean temperature, and the want of knowledge whether or not the same kind of barley is cultivated in all the places, will allow us to expect.

The general result as above detailed is thus generalized by Dr.

hydrangea become blue in a soil impregnated with carbonate of iron, and that all the brilliant spectacle of vegetable colors disappear either in consequence of accidents or on the approach of death. In connection with this subject it is curious to notice, he remarks, that the discoloration of plants is often determined by the same agents as in other cases produce color, and that certain organs which have no color whilst alive when dead have a decided tint! If the light be too powerful, a discoloration takes place in the plant. Cultivators of tulips with a knowledge of this fact place their flowers under a tent in order to preserve them from the effect of the direct action of the sun, which is known to alter their colors. The organic series of plants exhibit the same phenomenon.

Draper: "Every cultivated plant requires a certain quantity of heat for its development, but it is the same thing whether this heat is distributed over a shorter or longer space of time, so that certain limits are not exceeded; for where the mean temperature sinks below 36° 24', or where it rises above 71° 36', barley will no longer ripen. Consequently, to define accurately the conditions of temperature which a plant requires to maintain it in a flourishing condition, we must state within what limits its period of vegetation may vary, and what quantity of heat it requires. This was first observed by Boussingault."—"Human Physiology."

Most aquatic plants, like human beings, are said to acquire in death a whitish hue. Seaweeds of a most brilliant hue or green color become white when they die. The same change is observed in the fresh-water confervae.

Although the parts of plants which originally are white or black become more or less colored when exposed to the action of light, yet organs, once colored, do not in reality lose their hues when kept in darkness; if they sometimes appear to do so it is owing to this—that if half developed leaves are placed in the dark, they grow larger, and the green matter which colored them, being diluted by water and spread over a greater space, appears to be paler without being itself less colored.*

When speaking of the influence of the sun's light on plants, Dr. Draper says: "Through this agent the decomposition of carbonic acid is effected, and the plant obtains from the air the carbon it requires, out of which its solid structures are for the most part built. The rapidity with which the reduction of the carbonic acid takes place depends upon the brilliancy of the light, and the amount of

---

* On Botany. By Dr. Lindley.

carbon thus obtained upon that condition and the time of exposure conjointly. The amount of light received from the sun in any locality depends in a general way, as does the heat, upon the latitudes, but in both cases a multitude of disturbing agencies intervene. Variations of moisture control the supply of light by permitting a translucency, or establishing its opposite, a cloudiness or murkiness of the air. Other meteorological causes, as, for example, winds, by condensing or removing moisture, act in like manner; so also do astronomical conditions, especially by influencing the relative length of the day and night; for, as we advance toward the pole, the summer sun is above the horizon longer and longer. In Northern Europe, during the month of June, the sun never sets, but remains all night, if night it can be called, above the horizon; and, as Berzelius well remarks, "Under the influence of this midnight sun of the North, the life of plants runs through the same cycle of change in six weeks which it takes four or five months to accomplish in beautiful Italy."

In connection with this subject it is of great interest to consider not only the effect of the sun on the color of man, animals, and plants, but its influence on the varied phases of insect life. Their maximum degree of vital manifestation as well as their variety, richness of hue, etc., are found in countries where the solar beam attains its highest point of development.

The rate of increase of insect life, in proceeding from either pole to the equator, is said to be very various in different longitudes. Their numbers are small in both the polar regions—more abundant in Tasmania and New South Wales—more so in Southern and Western Africa, Colombia, and a maximum in Brazil; but North America has fewer species than Europe in the same latitude; and Asia is comparatively poor in species, in proportion to its great extent. The horrors of insect annoyance in the swamps of the great rivers of tropical America are vividly described by Humboldt.

The air is one dense cloud of poisonous insects to the height of twenty feet. In Brazil the vivid colors and metallic brilliancy of many of the beetles are extraordinary.

Among the more remarkable varieties of insect life deserving special mention, are first, the bees and ants. Of the former each country has peculiar species; but it is singular that the honey-bee of

North America has been introduced from Europe.

The ants, of which the species are almost innumerable, are found chiefly in hot and dry climates, and are, perhaps, of all insects, the most remarkable in their habits. The termite of tropical Africa builds pyramidal nests ten or twelve feet in height, hollowed into chambers of elaborate structure. The white ant of India devours everything of animal and vegetable origin, ascending by covered galleries (for they cannot bear the light) to the sap of furniture, beams, etc. But perhaps the most singular species of all is that of the parasol ants of Trinidad in the West Indies, which walk in long procession, each one carrying a cut leaf over its head, as a parasol in the sun, and these they deposit in holes ten or twelve feet deep under ground, apparently with no other object than to form a comfortable nest for a species of white snake, which is invariably found coiled up among them on digging out the deposit.* The scorpion extends in Europe to the north coasts of the Mediterranean, but is more abundant in Africa, both

---

* Mrs. Carmichael's "Domestic Manners, etc. etc., in the West Indies," ii. 327.---Mary Somerville's "Physical Geography," from whose works these facts are quoted.

North and South, where its bite has the singular peculiarity, that, although excessively painful on the first occasion of its infliction, and even dangerous to life, the constitution becomes hardened to it, the suffering is less on every subsequent occasion, and at length comes to be little regarded. Brazil produces a scorpion six inches in length. The locust, one of the most formidable scourges in countries infested by it, migrates in such masses as to darken the air for successive days, and when driven into the sea is sometimes thrown up as banks on the shore, poisoning the air by their decomposition for many miles in length. They are frequent in Syria and Barbary, whence they occasionally migrate to Italy, and during the summer of 1858 several species were taken in the country, and one or two in London. They are even said to cross the Mozambique Channel from the African Coast to Madagascar.*

This section of my subject would be imperfectly elaborated if I were to omit all reference to the marked effect produced by the chemical influence of light in modifying the active principle of certain

---

* As quoted by Sir John Herschel: "Physical Geography."

medicinal drugs.* The phenomena referred to are of deep practical importance, not only to the pharmaceutical chemist but to the physician. If the powdered leaves of hemlock, aconite, foxglove, or henbane, are exposed for any length of time to a strong light, their color becomes changed, "passing," says Mr. Hunt, "slowly from a green into a slaty gray, and ultimately into a dirty yellow." A decomposition also takes place by which the medicinal activity of the plant is seriously altered if not made altogether inert.

Hence the importance of preserving the previously mentioned pharmaceutical preparations in well-covered bottles, thus carefully excluding them from the influence of the sun's rays.

By what chemical or physiological laws are we to explain the curious facts referred to by Mr. Hunt, when alluding to the effect of light in modifying the potency of some vegetable tinctures, and par-

---

* Dr. Draper performed some interesting experiments on the production of hydrochloric acid by the direct union of chlorine and hydrogen under the influence of light, both artificial and solar, and also on the decomposition of peroxalate of iron, from which the carbonic acid is readily disengaged, by which he established that the primary condition essential for the chemical action of light is the absorption of some of its rays.

ticular medicinal powders? Pulverized rhubarb and ginger adhere with considerable firmness to the sides of the bottles exposed to the light, whereas the sides in shadow are left clear. If camphor is kept in a bottle, crystals will be formed on the glass upon which the light falls. If that side is turned from the light the crystals will be gradually removed, and again be deposited on those parts upon which the rays of light first impinge.

Hydrocyanic acid is supposed to undergo decomposition if exposed to the action of light; hence it is preserved in blue-colored bottles.

Gum guaiacum is well known to turn green under the prolonged influence of the solar light. Dr. Wollaston took two specimens of paper colored with a yellow solution of this gum in alcohol, and exposed one of them to air and sunshine, the other to air in the dark. The former was turned perceptibly green in five minutes, and the change was complete in a few hours, while the latter was not discolored after the lapse of many months.

In connection with this subject, I have to consider in detail the exceedingly interesting experiments which have been made with a view of ascertaining the modifications that take place in the vital and chemical influence of light when passed through media of varied tints. Apart from the colored rays observed when the spectrum is subjected to the effects of some absorbing medium, there is an agent which is said to permeate the glass or fluid, which has a decided *thermic* influence. These phenomena are recorded with philosophic exactness by Mr. Hunt. The following is a summary of the facts he has conclusively established in relation to this subject: the highest degree of temperature is not obtained behind red media, but a yellow or orange tint; in fact, the maximum degree of temperature is found behind a colorless fluid. Red glasses and fluids absorb a larger quantity of the heat rays than any others excepting black ones. The effect of the rays of light passing through the different colored media on the germination of plants has been beautifully illustrated.

The following is the order in which the colors stand, when viewed in relation to their vital or germinating effects: 1, orange; 2, red; 3, ruby; 4, yellow; 5, blue; 6, green. "The roots of tulips," says Mr. Hunt, "under the orange glass, developed the cotyledons a week earlier than those under the yellow, blue, or green glasses. The greatest progress in germination was made by the tulips under the yellow and orange glasses; but the leaves under each of these were by no means healthy, particularly under the yellow glass, but which had a singular delicate appearance, being of a very light green color, and covered with a most delicate white bloom."

Under the influence of the color previously referred to, the leaf stalks of the tulips shot up remarkably long, and were in both cases white. Under the orange glass, a small flower-bud appeared in the plant; it, however, soon perished with the plant itself. Under the yellow glass no buds appeared, and the vitality of the plant soon failed. Tulips exposed to the influence of light passing through ruby and red glasses shot up in a single lobe. The duration of the life of the plant did not extend over three or four weeks, and it did not rise over two or three inches above the soil. Underneath the green glass the plants grew slowly but strongly. The stem was long, having a small leaf at the end, not more than two-thirds of an inch in diameter. The flower-buds generated under these circumstances never could be made to blossom, notwithstanding the greatest care and attention was bestowed upon them. The attempt to develop the bud appeared to exhaust the vitality of the plant, and it soon died. Under blue glasses the roots germinated less quickly than in the open ground, but exhibited a greater degree of vitality. Somewhat similar phenomena were observed with regard to the germination of seeds when exposed to the influence of light passing through various colored media, solid or fluid in their nature. Dr. Draper found that under the influence of the bright sun of Virginia, plants have grown well in the light which has been made to permeate through a considerable thickness of intense yellow solution.

When considering the chemical and physiological influence of light it is impossible to overlook the important results obtained by Professors Bunsen and Kirchhoff by the aid of the spectroscope, or analysis of the sunlight by the prism. When this instrument is applied to the investigation of a solar beam it is found to be traversed by a number of vertical dark lines ("Fraunhofer's lines," as they are designated).* The sunlight is thus proved

---

* In order to explain the occurrence of the dark lines in the solar spectrum, we must assume that the solar atmosphere incloses a luminous nucleus, producing a continuous spectrum, the brightness of which exceeds a certain limit. The most probable supto contain various metals. Among these are iron, nickel, sodium, calcium, magnesium, chromium, and in small quantities, barium, copper, and zinc. Gold, silver, mercury, aluminium, cadmium, tin, lead, antimony, arsenic, strontium, and lithium are not, according to the experiments of Professor Kirchhoff, found in the solar atmosphere. It is supposed that the rays of light which form the solar spectrum and present the interesting phenomena previously referred to have passed through the vapor of iron, and have thus suffered the absorption which the vapor of iron must exert. It

---

position which can be made respecting the sun's constitution is, that it consists of a solid or liquid nucleus, heated to a temperature of the brightest whiteness, surrounded by an atmosphere of somewhat lower temperature. This supposition is in accordance with Laplace's celebrated nebular theory respecting the formation of our planetary system. If the matter, now concentrated in the several heavenly bodies, existed in former times as an extended and continuous mass of vapor, by the contraction of which sun, planets, and moons have been formed, all these bodies must necessarily possess mainly the same constitution. Geology teaches us that the earth once existed in a state of fusion; and we are compelled to admit that the same state of things has occurred in the other members of our solar system. The amount of cooling which the various heavenly bodies have undergone, in accordance with the laws of radiation of heat, differs greatly, owing mainly to difference in their masses. Thus whilst the moon has become cooler than the earth, the temperature of the surface of the sun

has been maintained that these iron vapors are contained in the atmosphere of the sun or in that of the earth. "It is not easy to understand," says Professor Kirchhoff, "how our atmosphere can contain such a quantity of iron vapor as would produce the very distinct absorption lines which are seen in the solar spectrum. This supposition is rendered still less probable by the fact that these lines do not appreciably alter when the sun approaches the horizon.

"It does not, on the other hand, seem at all unlikely, owing to the high temperature which we

---

has not sunk below a white heat. Our terrestrial atmosphere, in which now so few elements are found, must have possessed, when the earth was in a state of fusion, a much more complicated composition, as it then contained all those substances which are volatile at a white heat. The solar atmosphere at this present time possesses a similar constitution. The idea that the sun is an incandescent body is so old, that we find it spoken of by the Greek philosophers. When the solar spots were first discovered, Galileo described them as being clouds floating in the gaseous atmosphere of the sun, appearing to us as dark spots on the bright body of the luminary. He says that if the earth were a self-luminous body, and viewed from a distance, it would present the same phenomena as we see in the sun.—"Researches on the Solar Spectrum and the Spectra of the Chemical Elements." By G. Kirchhoff, Professor of Physics in the University of Heidelberg. Tanslated by Henry E. Roscoe, B. A., Professor of Chemistry, Owen's College, Manchester. London. 1862.

must suppose the sun's atmosphere to possess, that such vapors should be present in it. Hence the observations of the solar spectrum appear to prove the presence of iron vapor in the solar atmosphere with as great a degree of certainty as can be attained in any question of natural science".*

The important results thus obtained regarding the properties of light by means of the solar spectrum, will in all probability tend greatly to elucidate the physical constitution of the sun, the nebular theory respecting the formation of the planetary system, and the geological character of the earth; but into the consideration of the various hypotheses suggested to explain the phenomena relating to these interesting philosophical points, it would be foreign to my purpose to enter.

When addressing myself to the hygienic influence of light, I purpose again recurring to Professor Bunsen and Kirchhoff's discovery, for the purpose of ascertaining to what extent the development of the red-blood cell and the iron found in the general circulation depend upon the mechanical or chemical effect of the solar beam (contain-

---

* "Researches on the Solar Spectrum," etc. By Professor Kirchhoff.

ing in its composition this metal) upon the portions of the body exposed to its operation.

It is not my intention to analyze the morbid effects of solar light in the generation of specific diseases. I leave the consideration of this portion of the subject in the able hands of Sir James Ranald Martin, and other well-known writers on the Diseases of Tropical Climates.

That a prolonged exposure to the intense rays of the sun injuriously affects the health is generally admitted. Inflammation and congestion of the brain; engorgement, inflammation, enlargement, and torpidity of the liver; intermittent, remittent, gastric, yellow, and congestive fevers; dysentery, and cholera, are among the principal maladies incidental to a protracted residence in the tropical zone.

Fatal attacks of what in India is termed "heat apoplexy," or "coup-de-soleil," are common in various parts of that country, as the effect of an indiscreet exposure to the vertical rays of the sun.

PART II.

THE LUNAR RAY.

I WOULD refer briefly to the ancient opinions respecting the influence of the moon. From the earliest periods of antiquity, the idea generally prevailed, not only that the moon exercised a specific effect in the production and modification of disease, mental and bodily, but played a prominent and important part in the development of the character of nations, and in determining the destinies of the human race. Among the ancients the moon was viewed as an object of superstitious regard. They held her in great religious veneration, considering her influence superior even to that of the sun; in fact, they worshipped her as a Deity. The new moons, or the first days of the month, were kept with great pomp and ceremony as national festivals. The people were obliged to rest on those days. The feast of new moons was a miniature of the feasts of the prophets. Eclipses, whether of the sun or moon, were looked upon as ēvidences of Divine displeasure. The Greeks consulted the different phases of the moon before contracting marriage, and the full moon, or the times of conjunction of sun and moon, were considered the most favorable periods for celebrating the ceremony, in consequence of the impression that the reproductive functions were under lunar influence.

" This connection of the moon," says Dr. Laycock, "with the measure of time seems to have brought that planet into relation with the religious rites of ancient nations, as the Egyptians and Jews; and, also, to have given origin (in part) to the *mythological* idea so extensively prevalent of a lunar influence on marriage and childbearing. Even the barbarous Greenlanders, as Egede informs us, believe in this superstitious notion. They absurdly imagine that the moon visits their wives now and then; and that staring long at it when at its full will make a maid pregnant! Among the ancient nations the general idea was, that the lunar influence varied according to the age of the moon. Bubastis, the Egyptian Diana, was not equally favorable to parturient females and their offspring in her different phases. Among the Jews the full moon was believed to be lucky, and the other phases disastrous."

" The full moon," says the Rabbi Abravanel, " is propitious to new-born children : but if the child be born in the increase or wane, the horns of that planet cause death ; or, if it survive, it is generally guilty of some enormous crime." The Greeks and Romans entertained a similar idea respecting the lunar phases. The general opinion seems to have been, that the moon was propitious in proportion as its luminous face was on the increase. The ancient Greeks considered the day of the full moon to be the best day for marriage. Euripides makes Agamemnon answer, when asked on what day he intends to be married,

"Οταν Σελήνης εὐτυχὴς ἔλθη κύκλος"*

Hesiod asserted that the fourth day of the moon was propitious, but the eighteenth was bad, especially to females. Aristotle maintained that the bodies of animals were cold in the decrease of the

---

* "Iphigenia." Act v. 717.

moon, and that the blood and humors are then put in motion, and to those revolutions he ascribes the various derangements peculiar to women.

Lucilius, the Roman satirist, says that oysters and echini fatten during lunar augmentation, which also, according to Gellius, enlarges the eyes of cats; and that onions throw out their buds in the decrease of the moon, and wither in her increase, which induced the people of Pelusium to avoid their use. Horace also notices the superiority of shell-fish during the moon's increase.

Pliny takes notice of the same fact. He also adds that the streaks on the livers of rats answer to the days of the moon's age; and that ants never work at the time of the lunar changes. He also informs us that the fourth day of the moon determines the prevalent wind of the month, and confirms the opinion of Aristotle that earthquakes generally occur about the new moon. Pliny asserts that the moon corrupts all dead carcases exposed to its rays, and produces drowsiness and stupor in those who sleep under her beams. He further contends that the moon is nourished by rivers, as the sun is fed by the sea. Galen asserted that all animals who are born when the moon is falciform, or at the half quarter, are weak, feeble, and short-lived ; whereas those who come into the world during the full moon are healthy, vigorous, and long-lived.

Lord Bacon adopted the notion of the ancients. He maintained that the moon developed heat, induced putrefaction, increased moisture, and excited the motion of the spirits.* Van Helmont affirmed that a wound inflicted during the period of moonlight is most difficult to heal, and boldly asserted, that if a frog be washed clean and tied to a stake under the rays of the moon in a cold winter's night, on the following morning the body will be found dissolved in a gelatinous substance bearing the shape of the reptile, and that coldness alone, *without the lunar action*, will never produce the same effect !

The Spartans considered the moon to have great influence, and no motive could induce them to enter upon an expedition, or march against the enemy, until the full of the moon. The Greeks and Romans believed that the moon presided over child-birth. The patricians of Rome wore the figure of a crescent upon their shoes, to distinguish

---

* It is recorded that this great philosopher always had a severe attack of syncope at the time of a lunar eclipse.

them from the inferior order of men. The crescent was called *lulula*. Herodotus records that when the Lacedæmonians visited Athens, after the battle of Marathon, they waited until the moon had passed its full before they continued their march.* The ancient alchymists attempted to localize planetary influences, maintaining that the *heart*, which represented, according to their physiological notions, the vital principle, was under the special protection of the *sun;* that the *brain* was regulated and controlled by the *moon:* that *Jupiter* presided over the *lungs*, *Mars* the *liver*, *Saturn* the *spleen;* that *Venus* took the *kidneys* under her kind control, and *Mercury* sat in judgment upon the reproductive functions. They were also of opinion that the morning regulated the blood, noon the bile, evening the atra-bile, and night the cold phlegmatic constitution.

It will appear by the previously recorded data that from the earliest periods in the history of the world the idea of the phenomena of organic life being subject to planetary control, was popular among enlightened and philosophic men.† The

---

* Erato. lxx.

† "A knowledge of the ancient and popular belief in Sidereal influence will enable us to explain many superstitions in physic;

following passage proves that the great Roman satirist believed in lunar influences :—

the custom, for instance, of administering cathartic medicines at stated periods and seasons, originated in an impression of their being more active at particular stages of the moon, or at certain conjunctions of the planets; a remnant of this superstition still exists to a considerable extent in Germany; and the practice of bleeding at 'spring and fall,' so long observed in this country, owed its existence to a similar belief. It was in consequence of the same superstition that the metals were first distinguished by the names and signs of the planets; and as the latter were supposed to hold dominion over time, so were astrologers led to believe that some, more than others, had an influence on certain days of the week; and, moreover, that they could impart to the corresponding metals considerable efficacy upon the particular days which were devoted to them; from the same belief, some bodies were only prepared on certain days in the year; the celebrated earth of Lemnos was, as Galen describes, periodically dug with great ceremony, and it continued for many ages to be highly esteemed for its virtues; even at this day, the pit in which the clay is found is annually opened with solemn rites by the priest, on the 6th day of August, six hours after sunrising, when a quantity is taken out, washed, dried, and then sealed with the Grand Signior's seal, and sent to Constantinople. Formerly it was death to open the pit or to seal the earth on any other day in the year. In the botanical history of the Middle Ages, as more especially developed in Macer's Herbal, there was not a plant of medicinal use that was not placed under the dominion of some planet, and must neither be gathered nor applied but with observances that savored of the most absurd superstition, and which we find were preserved as late as the seventeenth century by the astrological herbarists, Turner, Culpepper, and Lovel."—Dr. Paris's "Pharmacologia."

> "Ut mala quem scabies aut morbus regius urguet,
> Aut fanaticus error, et iracunda Diana,
> Vesanum tetigisse timent fugiuntque poëtam,
> Qui sapiunt."*

These notions have not been confined to classical regions, ancient authorities, or to the fanciful creations of the poet. They have existed among barbarous, uncivilized, and unlearned nations, who were profoundly ignorant of the views propounded by the astrologers of old, or by the medical writers, who had somewhat engrafted the study of medicine upon that of astrology and astronomy. In referring to the alliance which formerly obtained between the two sciences, it has been well observed by an able writer and close observer of nature, "that no judicious person can doubt that the application of astrology to medicine, though it was soon perverted and debased till it became a mere craft, originated in *actual observations of the connection between certain bodily affections and certain times and seasons.* Many, if not most, of the mischievous systems in physics and divinity have arisen from dim perception or erroneous apprehensions of some important truth, and not a few have orig-

---

* Hor. "Ars Poetica."

inated in the common error of drawing bold and hasty inferences from weak premises."*

That the theory of planetary influence should have been advocated in early times, and have found zealous supporters, not only among the illiterate, but among learned and scholastic men, need not excite surprise when we consider how easily susceptible of demonstration is the fact of the moon's powerful effect in producing that regular flux and reflux of the sea which we call tides. Astronomers having admitted that the moon was capable of producing this physical effect upon the waters of the ocean, it was not altogether unnatural that the notion should become not only a generally received, but a popular one, that the ebb and flow of the tides had a material influence over the bodily functions. The Spaniards imagine that all who die of chronic diseases breathe their last during the ebb. Southey says, that among the wonders of the isles and city of Cadiz, which the historian of that city Suares de Salazar enumerates, one is according to P. Labat, that the sick never die there while the tide is rising or at its height, but always during the ebb. He restricts the notion to the isle of Leon,

---

* Southey.

but implies that the effect was there believed to take place in diseases of all kinds acute as well as chronic. "Him fever," says the Negro in the West Indies, "shall go when the water come low; him always come not when the tide high." The popular notion among the Negroes appears to be that the ebb and flow of the tides are caused by a *"fever of the sea,"* which rages for six hours, and then intermits for as many more.

There exists in the writings of many able, truthful, and conscientious men a vast body of valuable and indisputable evidence in support of the theory of planetary influence. I subjoin the names of the principal authorities on the subject:—Ballonius, Ramazzini,* Joubertus,† Joannes Morellus,‡ Mead,§ Gemma,|| Paræus,¶ Dr. Nicolas Fontana,** Dr. Cullen,†† Dr. Balfour,‡‡ Dr. James

---

* De Constitutionibus trium sequentium annorum, 1692, 1693, 1694, in mutinense civitate et illius Ditione, Dissertatio; which essay will be found in the first volume of his Opera Omnia Medica et Physiologica.

† On Epidemics.   ‡ On Putrid Fever.

§ De Imperio Solis et Lunæ in corpora humana et Morbis inde oriundis.

|| On the Plague of 1575.   ¶ On the Plague.

** Osservazioni sopra le Malattie che attacano li Europei nei Climi caldi, etc. Livorno, 1781.

†† First Lines.

‡‡ Effects of Sol-Lunar Influence in Fevers. London, 1815.

Lind,* Dr. Jackson,† Dr. James M'Grigor,‡ Dr. James Gilchrist,§ Dr. James Johnson,‖ Dr. Liddell,¶ Dr. Diemerbroeck;** and in our own immediate epoch, Drs. Orton,†† Radcliffe, and Laycock on the periodicity connected with disease and associated with the vital phenomena‡‡—Sir James Ranald Martin, C. B., F. R. S.,§§ Dr. Milligan,‖‖ William Ramsay,¶¶ Dr. Prichard,* Arago,† and Dr. Lardner.‡ Many of the great medical authorities of antiquity were clearly of opinion that the celestial bodies exercised a marked influence upon the bodily and mental functions. Dr. Haslam asserts

---

* On Putrid Fevers.
† Treatise on the Connection of the New and Full Moon with the Invasion and Relapse of Fevers.—*London Medical Journal,* for 1787. Also, his Treatise on the Fever of Jamaica.
‡ Medical Sketches of an Expedition to Egypt.
§ On the Diseases of India.
‖ On the Diseases of Tropical Climates.
¶ On the Diseases of Tropical Climates.
** On the Plague.
†† On Cholera.
‡‡ Vols. ii. and iii., *Lancet,* 1842-3.
§§ On the Diseases of Tropical Climates. Second Edition.
‖‖ Curiosities of Medical Experience.
¶¶ Astrologia Restaurata.
* Analysis of the Egyptian Mythology.
† Meteorological Essays.
‡ On Lunar Influence.

that Hippocrates, whom he designates as a "philosopher and correct observer of natural phenomena," did not place any faith in the generally received notion respecting the influence of the moon. This is clearly an error. Hippocrates imbibed so strong a belief regarding the effects of the celestial bodies upon the vital manifestations that he expressly recommends no physician to be entrusted with the treatment of disease who was ignorant of astronomical science; and he expressly advises his son Thessalus, to study the science of numbers and geometry, affirming that the "rising and setting of the stars have great effect upon the distempers."*

The critical days, or *crises* as they were termed, were said to correspond with the interval between the moon's principal phases.† Galen adopted the Hippocratic notion. Hence the lunar periods were said by him to be connected with the exacerbation of particular diseases.

---

* Epist. ad Thessalum de aëre, aquis, et locis.

† The crises which Hippocrates describes by the words *imperfecte judicabantur*, were, according to Dr. Balfour, nothing more than *intermediate inter-lunar crises;* and those to which he applies the terms *perfecte judicabantur*, were *final inter-lunar* crises.

The doctrine of lunar influence has descended to modern times, and notwithstanding a section of the scientific world has altogether repudiated the idea, it has nevertheless found zealous advocates among learned men. Writers of admitted judgment and sagacity have been recognized in the ranks of those who support this theory. At the threshold of this important and interesting inquiry it will be well to pause and consider, why any number of men of science should exhibit a disposition to discountenance this notion of planetary influence? Dr. Orton endeavors to answer the question.

"The difficulty of explaining lunar influence appears to be the great obstacle which in modern times has stood in the way of the belief of its existence and general prevalence. The ancients, who less minutely scrutinized the chain which connects effects with remote causes, implicitly believed in the existence of this power, simply because they saw the coincidence of its effects and certain states of the heavenly bodies, although they knew not that these bodies in other respects exert a physical influence on the earth. But since the progress of science has enabled men to trace more distinctly the manner in which changes arise from and produce other changes, this empirical mode of reasoning has ceased to be satisfactory; and the improvement of philosophy seems, in some instances, to have actually operated as a barrier to its further progress, by furnishing negative arguments against the existence of causes which we are unable to connect by any satisfactory theory with their effects. Every occurrence in Nature has been attempted to be accounted for on rational and general principles, and it has been found much easier to deny than to explain the operation of the sol-lunar power. If, however, these principles were to be applied in all their extent to the other branches of medicine, they would strike at the very root of that imperfect science; for we know little more of the *modus operandi* by which ipecacuanha produces vomiting, or jalap produces purging, than we do of that by which the new or full moon produces attacks of intermittent fever, of mania, or epilepsy. We have the same kind of evidence of the agency of both these classes of causes; and after the proofs which have been adduced of sol-lunar influence, it would be nearly as preposterous to deny its existence—because we cannot account for it, because it does not produce its effects on all persons, or because the same occurrences frequently arise without its agency, as it would be to assert that a common dose of ipecacuanha or jalap will not produce vomiting or purging for precisely the same reasons. It does not, nevertheless, appear to be impossible to make some approach to the explanation of the nature of sol-lunar influence on known principles. It is proved, on the known laws of gravitation, that the various situations of the moon necessarily must have determinate effects on the atmosphere. Observations have shown that such is the case, and on these data considerable progress has already been made in the elucidation of this interesting subject.

" It appears to be very evident that sol-lunar influence is much more powerful within the tropics than in other parts of the world; and this may in some degree account for the little credit which it has met with; for little information, in comparison to the opportunities which are presented, has been conveyed from these countries to the native regions of philosophy. Dr. Balfour has indeed been impressed with all the importance of his subject, and even more than all; his situation and experience were such as to entitle his opinions to the highest attention, and he has given them to the world in the fullest manner; but he has failed in gaining a complete credit, probably from the dogmatical style which he has adopted, and from his having fallen into the error which is usually fatal to theorists—that of aiming at too much."*

Dr. Balfour's treatise will form the basis of some remarks when I come particularly to analyse the facts recorded by the different authorities relative to lunar influence in the production of disease. There has, I think, been a disposition to discourage of late years any minute, special, and scientific investigation of the facts recorded by men of veracity, on the presumption that the subject is altogether fanciful, visionary, and utopian. If the question has been seriously considered with a view to elicit truth, has the inquiry been calmly and dispassionately pursued, and that, too, by competent observers, possessed of the preliminary amount of mathematical, astronomical, and meteorological science indispensably necessary in order to arrive at anything like a satisfactory result or scientific conclusion? I much doubt the fact. In general conversation on the subject, the observation is often made, " Oh, I have not overlooked the

---

* Dr. Reginald Orton's " Essay on the Epidemic Cholera of India." p. 202. 1831.

study of the subject; I have been careful to observe whether the moon does really exercise any influence in modifying the type of disease, and have arrived at the conclusion that the notion is a puerile and fallacious one." But when the question is asked as to the *mode of investigation* which has been adopted, it will generally be found to have been loose and unscientific. With undoubtedly a sincere disposition to arrive at the truth, the method adopted by the inquirer has not been so sufficiently philosophic, logical, and exact as to entitle it to the respect of learned men. To establish the inconsistency displayed by writers on the subject, Dr. Orton cites passages from two standard works of scientific reference, relating to the subject of lunar light, in which the authors deny *in toto* its effects on the human organism. " The hypothesis of planetary influence," says one of the authorities, " has originated and passed by with the age of astrology."*

Another writer remarks that, "as the most accurate and sensible barometer is not affected by the various positions of the moon, it is not thought likely that the human body should be affected by

---

* Rees' " Encyclopædia." " Encyclopædia Britannica."

them." "But in the following page," says Dr. Orton, "*the writer furnishes a body of evidence to establish that the barometer has been found to be very remarkably affected by the various positions of the moon.*" It is not easy to reconcile such statements.

Before proceeding to analyse the facts cited by the authorities previously referred to, as illustrative of the influence of lunar light in the production of bodily disease, I would briefly direct attention to some of the well-known data regarding periodicity, as associated with the origin, progress, and type of disease. The theory of lunar influence is in a great measure based upon this well established law. The doctrine of periodicity, as exhibited in the phenomena of life, is not altogether of modern origin.* The ancients were too close and accurate

---

* Certain plants appear, in an inexplicable manner, to be influenced by the law of *periodicity* as well as by light. Dr. Balfour, when referring to this subject, observes that a plant accustomed to flower in daylight at a certain time will continue to expand its flowers at the wonted time, even when kept in a dark room. De Candolle made a series of experiments on the flowering of plants kept in darkness and in a cellar lighted by lamps. He found that the law of periodicity continued to operate for a considerable time, and that in artificial light some flowers opened, while others, such as species of convolvulus, still followed the clock hours in their opening and closing.

students and observers of nature to have overlooked the fact. The phenomena of menstruation were the subject of particular observation in all ages, and its singular and well-marked periodical character was attributed to the operation of causes acting independently of those organic laws supposed to regulate the special functions of life. This periodicity is observed in a large class of febrile affections, particularly in the intermittent, remittent, and bilious fevers of tropical climates, in the class of disease termed *neuroses*, in all spasmodic and convulsive disorders, particularly in epilepsy and its allied affections, in many forms of insanity, and in the diseases classed under the term *exanthemata*.

There is much in the recorded facts and observations embodied in the valuable treatises of Drs. Mead and Balfour, to strengthen the presumption that the periodicity referred to arises directly or indirectly from sol-lunar influence. Medical meteorology has not yet assumed the character and position of an exact and demonstrative science, and although I would concede much to those who have patiently considered this interesting branch of philosophic inquiry, I am in duty bound to pause before attributing too much power to those external agents (active I admit them to be) that are considered to regulate and control the great principle of life, either in its healthy or morbid manifestations. Can it not be demonstrated that the vital law regulating the phenomena of menstruation acts independently of certain external agencies? I repeat, is this fact not susceptible of proof? Until we are satisfied that this important uterine function is not dependent upon a special organic law inherent in or acting specifically upon the uterus itself, shall we not be travelling beyond the limits of a safe and logical induction, by assuming as an indisputable and demonstrable fact, that the phenomena to which I refer are the effect of lunar conditions, or dependent upon certain meteorological states of the atmosphere induced by the physical aspects of the moon?*

I now proceed with the historical analysis of the subject.

Dr. Mead's treatise† appeared soon after Sir Isaac Newton's immortal discoveries burst like a

---

* These periodical discharges are said to be more profuse in countries near the equator than towards the poles. Hippocrates notices this fact, and used it to explain the sterility of the women of Scythia.

† "De Imperio Solis et Lunæ in corpora humana et Morbis inde oriundis."

flood of dazzling light upon the world. Dr. Mead occupied a high position amongst the *literati* of Europe. His reputation as a scholar, physician, man of letters, and a lover and cultivator of science, was universally established. He was the intimate friend of Pope, Newton, and of Halley. He stood high in the estimation of foreign princes and kings, and the learned and scientific men of all countries eagerly sought his acquaintance, and felt honored by his friendship. It is recorded in his biography that the King of Naples forwarded to Dr. Mead the two first volumes of Signor Bajardi's erudite work on the antiquities of Herculaneum, paying him the compliment of asking in return a complete collection of his own works, and at the same time inviting him to his palace, for the purpose of showing him his valuable collection of Herculaneum antiquities. Considering the position of Dr. Mead, everything that fell from his pen was read with avidity, and his observations on all subjects were considered to be based upon a patient and accurate study of the great book of nature. His work previously referred to, was read with universal interest; and although it gave rise to much controversy, it nevertheless commanded the respect of his learned contemporaries. It was the first modern treatise on the subject, and proceeding from a physician of Mead's reputation, it at once formed the topic of general conversation and criticism. Such being the character of the work, I proceed briefly to analyse its contents.

Dr. Mead, in the prefatory part of his treatise, dwells much upon the importance of a previous acquaintance with the mathematical principles of natural philosophy, in order fully to comprehend the subject of lunar influences.

He then attempts to demonstrate, in the first place, that the sun and moon, in proportion as they approach near the earth, independently of their influence upon heat and moisture, must, at certain times, materially modify vital phenomena. The author, in the second place, cites facts illustrative of his theory, and then makes some suggestions in reference to the practical division of the subject.

Dr. Mead enters fully into the consideration of the effect of the moon on the winds, observing that the most boisterous seasons of the year occur about the vernal and autumnal equinox. It is a matter, he remarks, of common observation, that in the calmest weather there is some breeze at mid-day, at mid-night, and also at full sea—that is about the time the sun and moon arrive at the meridian, either over or under our hemisphere. Without entering more minutely into analysis of Dr. Mead's able and ingenious essay, his theory of sol-lunar influence may be thus briefly epitomized: According to Dr. Mead, the attraction of the sun and moon being increased at the *syzygies* (new and full moon), and the *perigees* (those situations in the moon's orbit in which she approaches nearest to the earth), and the passages over the equator, the weight of the atmosphere is consequently diminished, and it is rendered *mechanically unfit for respiration, and for supporting the due degree of pressure on the surface of the body.* Dr. Mead endeavors to establish, on Newtonian principles, that in all the situations in which the sun and moon have been found to produce their greatest effects in raising the tides, rarefying and disturbing the atmosphere, and in producing disease, their joint attraction for the earth, or certain parts of it, is greatest; and, on the contrary, where these effects are least evident, that these attractions are least. Dr. Mead maintains that the atmosphere is much more under lunar attraction than the ocean, owing to its greater height, which removes it further from the earth and nearer to the moon. Dr. Mead supposes that the influence of the moon is most visible in low conditions of vitality and in certain states of disease, and its effects are said to be more manifest on the nerve-force than on the blood, or any other of the animal fluids. I consider it, however, fair that Dr. Mead should, to a certain extent, be the exponent of his own views. I therefore make no apology for quoting, *in extenso*, two important passages from his treatise, having special reference to his theory on lunar influence :—

" It has been for a considerable time established that our atmosphere is a thin elastic fluid, one part of which gravitates upon another, and whose pressure is communicated every way in a sphere to any given part thereof. From hence it follows that if by any external cause the gravity of any one part should be diminished, the more heavy air would rush in from all sides around this part to restore the equilibrium which must of necessity be preserved in all fluids. Now this violent running in of the heavier air would certainly produce a wind, which is no more than a strong motion of the air in some determined direction. If, therefore, we can find any general cause that would, at these stated seasons which we have mentioned, diminish the weight or pressure of the atmosphere, we shall have the genuine reason of these periodical winds, and the necessary consequences thereof. The flux and reflux of the sea was a phenomenon too visible, too regular, and too much conducing to the subsistence of mankind, and all other animals, to be neglected by those who applied themselves to the study of nature. However, all their attempts to explain this admirable contrivance of infinite wisdom were unsuccessful till Sir Isaac Newton revealed to the world juster principles, and, by a truer philosophy than was formerly known, showed us how by the united or divided forces of the sun and moon, which are increased and lessened by several circumstances, all the varieties of the tides are accounted for. And since all the changes we have enumerated in the atmosphere do fall out at the same times when those happen in the ocean, and likewise whereas both the waters of the sea, and the air of our earth, are fluids subject in a great measure to the same laws of motion, it is plain that the rule of our great philosopher takes place here; viz., that natural effects of the same kind are to be attributed as much as possible to the same causes.* What difference that known property of the air, which is not in water, makes in the case, I shall

---

* Sir I. Newton's "Principia." p. 387.

show anon. Setting aside the consideration of that for the present, it is certain that, as the sea is, so must our air, twice every twenty-five hours, be raised upwards to a considerable height, by the attraction of the moon coming to the meridian; so that, instead of a spherical, it must form itself into a spheroidal figure, whose longest diameter, being produced, would pass through the moon. That the like raising must follow, as soon as the sun is in the meridian of any place either above or below the horizon ; and that the moon's power of producing this effect exceeds that of the sun in the proportion of four-and-a-half to one nearly. Moreover, that this elevation is greatest upon the new and full moons, because both sun and moon do then conspire in their attraction; least on the quarters, in that they then are drawing different ways, it is only the difference of their actions that produces this effect; lastly, that this intumescence will be of a middle degree at the time between the quarters and new and full moon. The different distances of the moon in her perigæum and apogæum likewise increase or diminish this power. Besides, the sun's lesser distance from the earth in winter is the reason that the greatest and least attraction of the air upwards more frequently happens at a little before the vernal and the autumnal equinox. And in places where the moon declines from the equator, the attraction is greater and less alternately, on account of the diurnal rotation of the earth on its axis.

"Whatever has been said on this head is no more than applying what Sir Isaac Newton has demonstrated of the sea to our atmosphere; and it is needless to show how necessarily those appearances just now mentioned of winds, at the stated times, must happen hereupon. It will be of more use to consider the proportion of the forces of the two luminaries upon the air to that which they have upon the waters of our globe, that it may the more plainly appear what influence the alterations hereby made must have upon the animal body."

Dr. Mead then proceeds to demonstrate how much more powerfully the moon influences the atmosphere than the sea, and that the tides in the air, from lunar attraction, are much greater than on those of the ocean; and, after considering the effect of certain unnatural states of the atmosphere upon the barometer, and then the connection between certain states of the barometer and special as well as epidemic diseases, he, in the subjoined passage, further developes his views as to the mechanical influence of certain conditions of the atmosphere on the respiratory organs :—

"It will not be difficult to show that these changes in our atmosphere at high water, new and full moon, the equinoxes, etc., must occasion some alterations in all animal bodies, and that from the following considerations :—

"1st.—All living creatures require air of a determined gravity, to perform respiration easily and with advantage, for it is by its weight chiefly that this fluid insinuates itself into the lungs. Now, the gravity, as we have proved, being lessened by these seasons, a smaller quantity than usual will insinuate itself; and this must be of smaller force to comminute the blood and forward its passage into the left ventricle of the heart, whence a slower circulation ensues, and the secretion of the nervous fluid is diminished.

"2nd.—This effect will be the more sure in that the elasticity of the atmosphere is likewise diminished. Air proper for respiration must be, not only heavy, but also elastic to a certain degree; for as this is by its weight forced into the cavity of the thorax in inspiration, so the muscles of the thorax and abdomen press it into the most minute ramifications of the bronchia in expiration ; where, the bending force being somewhat taken off, and springy bodies, when unbended, exerting their power every way in proportion to their pressures, the parts of the air push against all the sides of the vesiculæ and promote the passage of the blood. Therefore, the same things which cause any alterations in the property of the air will more or less disturb the animal motions. We have a convincing instance of all this in those who go to the top of high mountains ; for the air is there so pure (as they call it)—that is, thin—and wants so much of its gravity and elasticity, that they cannot take in a sufficient quantity of it to inflate the lungs, and therefore breathe with great difficulty.

" 3rd.—All the fluids in animals have in them a mixture of elastic aura, which, when set at liberty, shows its energy, and causes those intestine motions we observe in the blood and spirits, the excess of which is checked by the external ambient air, while these juices are contained in their proper vessels. Now, when the pressure of the atmosphere upon the surface of our body is diminished, the inward air in the vessels must necessarily be enabled to exert its force in proportion to the lessening of the gravity and elasticity of the outward ; hereupon the juices begin to ferment, change the union and cohesion of their parts, and stretch the vessels to such a degree as sometimes to burst the smallest of them. This is very plain in living creatures put into the receiver exhausted by the air-pump, which always first pant for breath, and then swell, as the air is more and more drawn out; their lungs at the same time contracting themselves, and falling so together as to be hardly discernible, especially in the lesser animals."*

Making due allowance for the obsolete terms used by Dr. Mead, as well as for the state of pathological and physiological science of his epoch the reader will be able to detect, in the language which he adopts to enunciate the theory of lunar influence, the germs of some great truths, which have subsequently been confirmed, in all quarters of the globe, by appeals to the great book of nature. Dr. Mead has undoubtedly laid himself open to the charge of attempting to prove too much; but are not all ardent and zealous cultivators of science exposed to the same imputation?

In the concluding part of the essay, Dr. Mead details a number of facts that have come under his

---

* "Esperienze dell' Accademia del Cimento." p. 118.

own as well as the observation of his contemporaries, demonstrative of lunar influence. Some of the cases cited appear to have a somewhat fabulous origin; but making every allowance for some trifling and natural exaggerations into which the author has fallen in his zealous endeavors to substantiate his pet theory, all who read his essay must admit that it is to a great extent based on a clear and accurate observation of facts, however loosely and inaccurately they may have, in a few instances, been recorded. It will be interesting, while glancing at the literary history of this subject, to refer to some of Dr. Mead's illustrations. Dr. Mead was physician to St. Thomas's hospital during the time of Queen Anne's wars with France, and whilst occupying this honorable position, a great number of wounded sailors were brought into the hospital. He observed that the moon's influence was visible on most of the men at that time under his care. He then cites a case, communicated to him by Dr. Pitcairne, of a patient, thirty years of age, who was subject to epistaxis, whose affection returned every year in March and September—that is, at the new moon near the vernal and autumnal equinoxes. Dr. Pitcairne's own case is referred to as a remarkable fact corroborative of lunar influence. In the month of February, 1687, whilst at a country seat near Edinburg, he was seized, at nine in the morning, the very hour of the new moon, with a violent hæmorrhage from the nose, accompanied with severe syncope. On the following day, on his return to town, he found that the barometer was lower at that very hour than either he or his friend Dr. Gregory, who kept the journal of the weather, had ever observed it; and that another friend of his, Mr. Cockburn, professor of philosophy, had died suddenly, at the same hour, from hæmorrhage from the lungs, and also that six of his patients were seized, *at the same time with various kinds of hæmorrhages*, all arising, it was supposed, from the effect of lunar influence on the condition of the barometer.

He refers to the case of a young man, of delicate habits who brought on an attack of hæmoptysis by making an effort beyond his strength. The hæmorrhage during eighteen months regularly recurred at the full of the moon. Two remarkable instances, illustrative of the same fact, are recorded in the "Philosophical Transactions."* The first is that of a young man, who, from his childhood

---

* Nos. 171 and 172.

till the twenty-fifth year of his age, discharged a small quantity of blood from the corner of the thumb-nail of his left hand, every time the moon came to its full. The other is the case of a patient, who from the fifty-third to the fifty-fifth year of his age, had a periodical evacuation of blood from the extremity of the forefinger of his right hand.

Baglivi cites the case of a student at Rome who had a fistulous ulcer of the abdomen which appeared to have some connection with the colon. This discharged so abundantly on the increase, and so little on the decrease of the moon, that it served him as a perfect index of the periods and quadratures of that planet. Nephritic attacks, he says, frequently follow the course of lunar attraction.

Tulpius relates that Mr. Ainsworth, an English clergyman at Amsterdam, constantly suffered from an attack of the gravel, accompanied with suppression of flatus, at the full of the moon, which continued until she had made some progress in waning. Van Helmont mentions this influence of the moon on asthma; and Sir John Fluyer, who, from being personally afflicted with this disease, had more occasion to attend to its phenomena than most people, asserts that paroxysms of asthma are always most severe at certain periods of the moon, and commonly recur with the change. Still more extraordinary effects are attributed to the lunar influence. The celebrated Kerckringius, in his Anatomical Observations, mentions the case of a young lady who became plump and handsome with the increase of the moon, but who completely changed with the decrease of that planet. About the change, she became so disfigured and haggard that she secluded herself from all society for some days. Mead also refers to a lady, whose countenance always developed itself with the increase of the moon, so that the *éclat* of her charms always depended upon that planet.

Having given the preceding sketch of Dr. Mead's essay, I now proceed to analyze Dr. Balfour's treatise, the second work of any importance specially devoted to this subject.

Dr. Francis Balfour's first dissertation appeared in Calcutta, in 1784.* In 1790, in a "Treatise on Putrid Intestinal Remitting Fevers," published at Edingburgh, the periodical return of febrile paroxysms and their coincidence with the periodi-

---

* "Treatise on the Influence of the Moon in Fevers." This was subsequently reprinted in England, and inserted in Dr. Duncan's "Medical Commentaries."

cal revolutions and remissions of sol-lunar power, which constitutes the foundation and proof of this theory, was investigated, described, and illustrated by two different plates, exhibiting a synoptical view of the whole system. The first part of that treatise is a regular logical synthesis, arising from facts observed and collected by himself to the discovery of certain prevailing tendencies in nature, and thence to axioms or general laws. The second part is an analysis, in which these axioms or laws are employed to explain some of the most remarkable phenomena of fevers. The third part is an application of the principles of this theory to form general rules for practice.

This physician appears to have devoted great attention to the consideration of this subtle and disputed point in science, and, with a view to its satisfactory elucidation placed himself in communication with all the medical men of note resident in our Indian presidencies, elicting from them the result of their observations on the subject. Dr. Balfour maintains, that every type of fever prevalent in India, is, in a remarkable manner, affected by the revolutions of the moon. Whatever may be the form of fever, he says that he has invariably observed that its first attack is on one of the three days which immediately precede or follow the full or the change of the moon, so that the connexion which prevailed between the attack of the disease and the moon at or during the time referred to, was most remarkable. Relapses in cases of fever are also said frequently to occur at such times. Dr. Balfour has observed for a period of fourteen years, this tendency to relapse at the lunar full and and change; and, in particular cases, he was able to prognosticate the return of the fever at these periods with almost as much confidence as he could foretell the moon's revolution itself. Putrid, nervous, and rheumatic fevers of India are, according to Balfour, equally under the influence of the moon. In attempting to explain these phenomena, Dr. Balfour says, that along with the full and change of the moon there is constantly recurring some uncommon or adventitious state or quality in the air which increases fever and disposes to an unfavorable termination or crisis ; and that along with the intervals there is constantly recurring a state of quality in the air opposite to the former, which does not excite but diminishes fever and disposes to a favorable crisis.* Dr. Balfour

---

* It will be well to state what Dr. Balfour means by a crisis. He defines it to be " favorable changes which never fail to take

has collected a vast body of valuable evidence in support of his lunar theory, establishing beyond all dispute that in tropical climates the regular diurnal and septenary changes observed in the character of the fevers of India are coincident and correspondent with periodical sol-lunar conditions.

In the year 1783-4, Dr. Balfour had for many months the charge of a regiment of sepoys, in Cooch Behar, immediately under the vast range of mountains which separate the northern part of Bengal from Bootan. The prevalent diseases were fevers, or "fluxes" attended with fevers. During the first month four hundred men were invalided. The greater part, however, of these cases were convalescent in the course of the eight days that intervened between the full and change of the moon; but

---

place, in some degree or other, at the time of their *transition* from the lunar period in the inter-lunar interval, and generally on the first morning intermeridional interval after it; at which juncture the maturity of the critical disposition concurs with the periodical decline of sol-lunar influence in bringing them about; and they are distinguished by one or more of the following symptoms—viz. a sediment, or particular turbid appearance, of the renal secretion; a more free and natural perspiration; spontaneous evacuations; and cleaner, moister, and softer tongue, with a more free and natural discharge of saliva, a more loose and copious expectoration, and a free discharge of bile, which seems to disappear and be suppressed in the course of the fever," etc.

during the remaining months of his stay in that district, the diseases previously mentioned increased to almost double their extent at every full and change of the moon, falling down again to their former standard during the eight days which intervened between these two periods. With regard to small-pox occurring in India, Dr. Balfour expresses himself as perfectly satisfied that the full and change of the moon interfered with the eruption, and increased the accompanying fever to a dangerous degree.

Mr. Francis Day, of the Madras army, thus generalizes Dr. Balfour's observations:

1. That the influence of the moon is less apparent in Madras than in Bengal, but may be traced over every portion of our Eastern possessions.

2. That the lunar influence is thus exerted:— The first attack of fever almost invariably commences on one of those days preceding the full or new moon, or on one of those three which immediately follow them, but that the last three are the most violent in their effects.

3. That the new moon is more injurious than the full.

4. That during these times the most severe as well as the greatest number of cases take place, but

that when they occur at other periods they are less severe, and of short duration.

5. That these laws are as applicable to relapses as they are to primary attacks; so much so, that the author was often able to prognosticate the return of the fever at these periods with almost as much certainty as he could foretell the revolution (of the moon) itself.

Mr. Day epitomises the results of carefully prepared statistical observations as follows:—

1. That no decided preponderance in the admissions for malarious fever is observed at the time of the new moon.

2. That a decided preponderance is observed at the time of the full moon.

3. That more admissions occur in the three days preceding the full moon than in the three days subsequent to those changes.

4. That a slight increase in the admissions may be present about the first and third lunar quarters.

5. That the cases admitted at the time of the new moon are generally slightly more severe than the average admissions.

6. That the cases admitted at the time of full moon are much more severe than the average admissions.

7. That at times increased severity is also apparent at the first and third lunar quarters.

8. That the cases admitted during the three days preceding these changes are more severe than those admitted in the three subsequent ones.

From the entire series of cases, Mr. Day arrives at the probability that there is a sol-lunar influence which is greater in the equinoctial period than in the respective equinoctial intervals, and considerably more so in the autumn than in the vernal equinoctial period; that this force is greater at the full and new moon than at the intervals, and much more so at the full than at the new; that it is greater during the meridional period than at the intermeridional intervals, and much more so at the diurno-meridional than at the nocturno-meridional periods.*

The influence of the moon on the functions of life has been made the subject of observation and speculation in every part of India. The physiological and pathological effects of lunar light have been universally acknowledged by all medical men

---

* Quoted from Sir J. Ranald Martin's valuable work on the " Influence of Tropical Climates in producing the Acute Endemic Diseases of Europeans." London, 1861.

practising in tropical climates. The natives of India are taught to believe in lunar influence from early infancy. In the northern latitudes the effects of the moon's rays are said to be less sensibly felt than in India. In the latter country, those suffering from attacks of intermittent fever are often able to predict, by watching the phases of the moon, the accession of the disease. Balfour maintains that the fact of diseases appearing during every day of the month is no legitimate argument against lunar influence.

"The human body," he says, "is subject to alterations from a thousand external physical circumstances as well as from many internal moral affections. These lay the foundation of disease at every period of life, but they do not overthrow the evidence of lunar influence, although they are apt to mislead with regard to effects that depend on that alone. The human body is affected in a remarkable manner by the changes of the moon, I am perfectly convinced, although I cannot constantly pretend to see the operation of the general law, nor to account at all times for its perturbation, and agree in thinking that an attention to the power of the moon is highly necessary to the medical practitioner in India."

"It is a fact," says Dr. Orton, "which has been universally observed, particularly in tropical climates, that the moon has a great influence on the weather; the full and change tending to produce rain and storms, and the quarters being more frequently attended by fine weather." This is so well ascertained, and so thoroughly believed, at least in India, that it is nearly superfluous to adduce arguments or instances in support of it. On every side, then, we perceive the intimate connection which exists between the three series of phenomena which have been noticed—the great lunar periods, disturbed states of the atmosphere, and the attacks of the epidemic. It will also be proved that the other principal circumstance which has been supposed to attend the prevalence of cholera, the depression of the barometer, is likewise produced by the new and full and moon. Dr. Orton says, "Sollunar influence is, doubtless, but one of the causes producing the state of the atmosphere which gives rise to cholera; and I have no doubt that the disease will often be found to make its appearance when the disturbing power of the sun and moon is least, and to subside when that power is at its height. General exacerbations of other epidemics, as well as of cholera, will usually be found to correspond to the moon's syzygies, and the remissions of the quarters."

Dr. Kennedy bears testimony, in his work on Epidemic Cholera, to the influence of the moon. He observes, " that the constitution here (India), both native and denizen, is assuredly under lunar influence, or what is the same thing, under the influence of the changes of weather which invariably accompany the changes of the planet."

Diemerbroeck, in his well-known treatise on the Plague,* when speaking of the epidemic of 1636, says: "Two or three days before and after the new and full moon the disease was more violent; more persons were seized at these times than at others, and those who were then seized almost all died in a very few hours. Nescio quâ virium labefactione oppressi." In the dedication prefixed to this treatise, which is addressed to the prætor and consuls and the whole senate at Utrecht, he thus describes the nature of his own situation, the opportunities he had of acquiring a knowledge of the disease, and his object in publishing the work :—

" As in all well-constituted states it is the duty

---

* "Isabrandi Diemerbroeck Montferto Trajectini, antehac Noviomagi, nunc Ultrajecti Medici, de Peste." Libri Quatuor Dissertatio, &c. Arenaci, 1646.

of every one to contribute his advice and assistance for the public safety, that by their unanimous concurrence the present as well as impending evils of the state may be averted and repelled, I conceive that I should not act improperly if, concerning this plague, of all diseases the most cruel, and more destructful than an enemy, I, too, should offer some salutary advice toward the discovery of its hidden nature, together with some more certain method of curing it. For, as in warfare, none can so well elude the designs of the enemy, or repel his attacks, as one who has had experience in the art of war, so none can more effectually resist this cruel disease than one who has intrepidly opposed himself to its fury. This I did a long time ago, not only in the year 1633, when a most violent pestilential fever, the forerunner of this plague, afflicted most grievously the whole province of Gelderland, and principally the city of Nimeguen, where I was ordinary physician, and threw upon me so great a load of practice as hardly allowed me to take sustenance, but likewise, in 1636 and 1637, when the true plague raged so violently among the people of Nimeguen, and so great a number of sick was thrown upon my hands as to give me no rest or repose. Having at that time, with great danger, and at the risk of my life, investigated most inquisitively the nature of this most dreadful enemy, I now make public his *Portrait*, delineated in this book, for the safety of all."

The same writer asserts that during the epidemic fever which raged in Italy in 1693, patients died in great numbers on the 21st of January, at the periods of the lunar eclipse. But as Dr. Lardner observes, when recording the fact, it may be objected that the patients who then died in such numbers at the moment of the eclipse might have had their imaginations highly excited, and their fears wrought upon, by the approach of that event, if popular opinion invested it with danger. That such an impression was likely to prevail is evident from the facts which have been recorded. In 1654 at the time of a solar eclipse, such was the strong opinion entertained on this subject, that patients in considerable numbers were ordered by their physicians to be shut up in chambers well closed, warmed, and perfumed, with the view of escaping the injurious influence of the eclipse. The consternation that prevailed amongst all classes was very great, and such crowds rushed to the confessional, that the ecclesiastics found it impossible to exercise their spiritual vocations.

The late Dr. James Johnson observes,* when alluding to the fevers of India, however sceptical professional men in Europe may be in regard to planetary influence in fevers, &c., it is too plainly perceptible between the tropics to admit of a doubt. " I have, " he continues, "not only observed it in others, but have felt it in my own person in India when laboring under the effects of obstructed liver. That this influence predisposes to, or exacerbates, the paroxysms of fever in India and other tropical climates is incontestably proved by daily observation."

Dr. Scott, when speaking of the effect of sol-lunar light on the endemic and epidemic diseases prevalent on the western side of India says, " The influence of the moon upon the human body in this part of India is observed by every medical practitioner. It is universally acknowledged by the doctors of all colors, castes, and countries. The people are taught to believe in the fact in their infancy, and as they grow up they acknowledge it from experience. I suppose in the northern latitudes this power of the moon is far less sensible than in India. We here universally think that this state of the weakly and diseased bodies

---

* " On the diseases of Tropical Climates."

is much influenced by the movements of the moon. Many people know the very day on which their intermittents will make their appearance; and every full and change increases the number of patients of every practitioner. It is no argument against this influence that diseases appear during any day of the month. The human body is subject to alterations from a thousand circumstances, and from many affeetions of the mind. These lay the foundation for disease at any period; but they do not overthrow the evidence of lunar influence, although they are apt to mislead with regard to that alone."

Mr. Hutton, writing to Dr. Balfour from Calcutta, says that he has been at some pains during a considerable practice of some years at Prince of Wales Island to observe the effects of the moon on prevailing diseases of that place. The disease peculiar to that island are of the intermitting and remitting kind, dysenteries, diarrhœas, liver complaints, and rheumatic affections. " I have," he adds, " generally found the violence of the symptoms (in the above mentioned diseases) considerably increased during the full and change of the moon. I have noticed relapses to occur from the same cause. By keeping these facts in view, I have been enabled to administer the medicine adapted to the different diseases with a greater degree of certainty and precision."

Dr. Moseley remarks that the greater hæmorrhages from the lungs or those of plethora, like all periodical attacks of this kind (undisturbed in their natural course by peculiar circumstance), obey the influence of the moon. Of this, he says, he has had many proofs. That there are not more authenticated by others is owing, he believes, to the theory on which the fact depends not being sufficiently known to prevent the result escaping unnoticed. In another portion of his work he remarks that most of the patients whom he had attended in the spring of the year 1777, during attacks of fever, were much affected in the head at every new and full moon. He refers to the case of a man who had a severe attack of hæmoptysis always at the moon's full. When speaking of the mode of treating these hæmorrhagic conditions, he advises the physician to be watchful in every case of the kind when the moon's influence was considered to be greatest on the earth. He cites the history of a gentleman who suffered from hæmorrhage of the lungs who was advised to leave England during the winter and to reside in the south of France. Whilst there his attacks eame on periodically, *obeying faith-*

fully the principal changes of the moon. Dr. Moseley considers this to be one of the most decisive examples of lunar influence recorded in medical history. The following particulars of his illness deserve attentive consideration.

On February 14, 1786, when near *Toulon*, hæmorrhage came on; the moon was at its full on the preceding day. On February 29, when at *Aix* in *Provence*, he had another attack. There was a new moon on the 28th. The moon was again at its full on the 13th of April, and on the 15th the patient had another attack of hæmoptysis. A new moon appeared on the 28th of the same month, and on the 29th, when at *Tain* upon the *Rhone*, he had a relapse. At *Chalons*, in Burgundy, there was a full moon on the 13th of May, and on the 14th his hæmorrhage returned. At *Dijon*, June 11, when the moon was at its full, he had another attack. On July 11, at *Paris*, the moon was again at its full. At this lunar period the hæmorrhage returned. Again when at Yarmouth in the Isle of Wight, on August 9, the moon was then at its full. The hæmoptysis returned. Dr. Moseley alludes to the remarkable fact that the last three attacks of hæmorrhage from the lungs came on *at the instant the moon appeared above the horizon.**

---

* Dr. Moseley has pursued, in a few particulars, these investi-

Nicholas Fontana says, " That the influence of the moon is very visible in almost all febrile cases. I had a good opportunity of observing this fact in my first voyage to this country (Barrackpore), in the year 1777, and subsequently on board the Liverpool East Indiaman. The ship was driven on a bank in the river Massuma, on the east coast of Africa, at a place called Delagoa Bay. The crew, through hard labor and the unwholesomeness of the place, suffered much from epidemic bilious fever. We had 60 or 70 patients ill out of the crew of 180. Every one was seized with illness, and several died. In the Gulf of Cambay, at the Nicobar Islands, at Kedgeree, in the River Ganges, Hooghly, and on our return to St. Jago, one of tho Cape de Verde Islands, where the ships went,

---

gations to an extreme point. I say so without intending for a moment to throw discredit on his general statements. He asserts as a positive fact that persons who live to an extreme age, invariably die at the new or at the full moon. After citing a number of instances illustrative of this opinion, he remarks, " Here we see the moon shines on all alike, making (as in death) no distinction of persons in her influence :"—

" ———— *æquo pulsat pede, pauperum tabernas*
Regumque turres."

Hor. lib. 1, 4.
—" A Treatise on Tropical Diseases." B. Moseley, M. D. 1792.

and remained a certain time in the course of the voyage, the crew were severely and repeatedly attacked with remittent and intermittent fevers.

"I observed, and was so fully persuaded of the effects of the moon's changes on them that I took notice of the fact in a work I published after my return from Italy. I did so with the view of cautioning succeeding surgeons coming from the Mediterranean. In that country the method adopted in our practice was by no means fitted to the treatment of the acute diseases which Europeans were liable to between the tropics."

"Daily practice," says a well-known army surgeon, Dr. Millingen, "shows that paroxysms of fevers and various other maladies are under planetary influence. The evening gun in our garrisons was often the signal of severe exacerbation in certain febrile cases, while the *réveillée* developed acute aggravation in others." Sydenham and Floyer remark that gout and asthma were usually ushered in after the first sleep; and that during the prevalence of the cholera the invasion of the disease was noticed to occur towards daybreak.

The authorities previously cited conclusively establish that lunar influence is not to be viewed as a mere myth, or as an utopian speculation. A host of intelligent writers and acute observers confirm the fact beyond all dispute. It will remain for me to consider, not only the evidence in favor of the lunar theory, but the arguments advanced against the hypothesis. It is only by closely investigating both sides of the question that the philosopher in search of truth will be enabled to arrive at a safe deduction.

Before considering the effect of the moon upon the mind in a state of aberration it will be necessary to revert to the morbid and physiological effect of lunar light upon the vegetable kingdom. This has long been the subject of observation and speculation. Many curious and apparently inexplicable facts are upon record illustrative of the phenomena. It will be well to refer to some of the more reliable data in connexion with this division of my subject; but before doing so, I would remark that the Druids of Gaul and Britain, who combined the office of physician with that of the priest, believing with the ancients that all vegetable productions, and particularly medicinal plants, were constantly under planetary influence, were instructed to gather the far-famed mistletoe with a golden knife when the moon was six days old.

It is a question whether the vervain of the ancients was similar to the plant which now bears that name. The appellation *Verbena* or *Sagmina* was given by the ancients to various plants used in religious ceremonies.

The vervain, held in such high repute by the Romans, was gathered after libations of honey and wine at the rising of the dog star, and with the left hand, and, thus collected, served for various sacerdotal and medicinal purposes. Its branches were employed to sweep the temples of Jupiter; it was used in exorcisms for sprinkling lustral water; and, moreover, it was said to cure fevers, the bite of venomous reptiles, and it was alleged to have the virtue of allaying discord, and appeasing other violent passions. This plant so gathered, was always carried in the hand by heralds when sent to sue for peace. It was called *Verbenarii*. When its benign powers were shed over the festive board mirth and good temper were said to prevail.

When speaking of the all-powerful influence of this plant, Pliny says—

" *Nulla herba Romana nobilitatis plus habet quam hierabotane.*" *

It is stated as a fact that if peas are sown in the increase of the moon they never cease blooming;

---

* Plinii, lib. 18, c. 44.

that if fruits and herbs are set during the wane of the moon, they are not so rich in flavor, nor so strong and healthy, as when planted during the increase. All vines should be pruned at the wane of the moon, says Sibley, the astrologist. He also asserts that pomegranates will live only as many years as the moon was days old when they were planted. In planting shrubs, if they are desired to shoot up straight and late, and to take little root, they are to be set when the moon is in an airy sign, and in increasing light. Flowers that are under the influence of the moon only, open their blossoms at night, whilst those that are peculiarly under the government of the sun, open every morning when it begins to rise, and close in the evening when he sinks below the horizon. M. Auguste de Saint-Hilaire states that in Brazil, cultivators plant during the decline of the moon, all vegetables whose roots are used as food; and, on the contrary, they sow during the increasing moon the sugar-cane, maize, rice, beans, etc., and in general those which bear the food upon their stocks and branches. Experiments, however, were made and reported by M. de Chanvalon, at Martinique, on vegetables planted at different times in the lunar month, and no appreciable difference in their qualities was discovered. There are some traces of a principle in the rule adopted by the South American agronomes, according to which they treat the two classes of plants distinguished by the production of fruit on their roots or on their branches differently; but there are none in the European aphorisms. The directions of Pliny are still more specific: he prescribes the time of the full moon for sowing beans, and that of the new moon for lentils. " Truly," says M. Arago, "we have need of a robust faith to admit without proof, that the moon, at the distance of 240,000 miles, shall, in one position, act advantageously upon the vegetation of beans, and that, on the opposite position, and at the same distance, she shall be propitious to lentils." The wise husbandman is said to prune his vines in obedience to certain phases of the planet. It is a maxim among gardeners that cabbages, and lettuces which are desired to shoot forth early, flowers which are to be double, trees which it is desired should produce early ripe fruit, should severally be sown, planted and pruned during the decrease of the moon; and, on the contrary, that trees expected to grow with vigor should be sown, planted, grafted, and pruned during the increase of the moon. These opinions Dr. Lardner considers to be altogether erroneous. The increase or decrease of the moon, he maintains, has no appreciable influence on the phenomena of vegetation; and the experiments and observations of several French agriculturists, and especially of M. Duhamel du Monceau, have, he observes, clearly established this fact.

Mantanari has referred to physical causes for an explanation of the alleged lunar influence upon plants. During the day, he says, the solar heat augments the quantity of sap which circulates in plants, by increasing the magnitude of the tubes through which the sap moves, while the cold of the night produces the opposite effect by contracting these tubes. Now, at the moment of sunset, if the moon be increasing, it will be above the horizon, and the warmth of its light would prolong the circulation of the sap ; but, during its decline, it will not rise for a considerable time after sunset, and the plants will be suddenly exposed to the unmitigated cold of the night, by which a sudden contraction of leaves and tubes will be produced, and the circulation of the sap as suddenly obstructed. This explanation does not satisfy Dr. Lardner, who remarks, that if it be admitted that the lunar rays possess any sensible calorific power, this reasoning might hold good, but it will have very little force when it is considered that the extreme change of temperature which can be produced by the lunar light does not amount to the thousandth part of a degree of the thermometer! Upon this point, however, philosophers are at variance. The lunar rays have, according to the experience of practical men, a decided *calorific* effect.

The gardeners of Paris assured Arago that in the months of April and May they found the leaves and buds of their plants, when exposed to the full moon on a clear night, *actually frozen, when the thermometer in the atmosphere was many degrees above freezing point.* He mentions these facts as proving that the moon's rays have a frigorific power, but that the largest speculums directed to the moon produced no such indications on a thermometer placed in their focus.*

Dr. Howard, of Baltimore, has affirmed that on placing a blackened upper ball of his differential thermometer in the focus of a thirteen-inch reflecting mirror opposed to the light of the full moon, the liquor sank, in half a minute, eight degrees! Though the surface of the full moon exposed to us must necessarily be very much heated,

---

* Ferguss. " Bull. Univ." 1827, p. 383.

says Sir J. Herschel, possibly to a degree much exceeding that of boiling water—yet we *feel* no heat from it, and even in the focus of large reflectors it fails to affect the thermometer. No doubt, therefore, its heat (conformably to what is observed of that of bodies heated below the point of luminosity) is much more readily absorbed in traversing transparent media than direct solar heat, and is extinguished in the upper regions of our atmosphere, never reaching the surface of the earth at all. Some probability is given to this by the tendency to disappearance of clouds under the full moon, a meteorological fact (for as such we think it fully entitled to rank), for which it is necessary to seek a cause, and for which no other rational explanation seems to offer.*

Cases of sudden death and profound coma are recorded as the effect of an improper and prolonged exposure to the intense light of the full moon. Sailors have been found dead on deck after sleeping under the moon's rays. It is also said that convulsions, apoplexy, epilepsy, and insanity, sensations of oppression in the head, inertness and heaviness of the senses, have arisen from the same

---

* Outlines of Astronomy, pp. 261, 2. 1849.

cause. Plutarch observes :—" Everbody knows that those who sleep abroad under the influence of the moon are not easily awakened, but seem stupid and senseless."* In India, death has occasionally been known to arise from what is termed a *coup de lune*, or stroke of the moon ; and in Egypt, blindness has often been produced in persons who have imprudently fallen asleep with their faces exposed to intense lunar light. Does not Edgar Allan Poe refer to this morbid effect of the moon's rays in the following passage ?—

> " 'Neath blue bell or streamer,
> Or tufted wild spray,
> That keeps from the dreamer
> The moonbeam away."†

Dr. Madden mentions that the Arabs attribute a morbid influence to the moon, and think it causes ophthalmia and catarrh. He thought there was some influence from it in the desert, beyond the common dampness of the night.‡

The questions that naturally occur to the mind

---

* Plut. " Symp." B. 3.

† " Al Aaraaf," Tycho Brahe suddenly discovered in the heavens a star which attained in a few days a brilliancy surpassing that of the planet Jupiter, and then as quickly disappeared ! This phenomenon gave origin to the poem referred to.

‡ " Travels in Turkey."

in reference to this interesting inquiry are, whether the morbid phenomena alleged to result from the moon's rays are dependent upon the *mere intensity of lunar light*, or are to be considered as the effect of *some specific influence in the nature of the light itself?* Let me consider the first question. It is an admitted fact that the light of the full moon is at least 300,000 times more feeble than that of the sun. According to Humboldt, the mean distance of the earth from the sun is 12,032 times greater than the earth's diameter, therefore 20,682,000 German or 82,728,000 English geographical miles. The mean distance of the moon from the earth is 51,800 German or 207,200 English geographical miles.*

It is said that the solar light reflected from the surface of the moon is in every zone fainter than the solar light reflected in the daytime from a white cloud.†

When estimating the intensity of lunar light, Sir John Herschel affirms, that in the southern hemi-

---

* Cosmos.

† Sir J. Herschel says that it appears to be placed beyond a doubt that the moon acts directly as a magnet on the earth's magnetism, producing periodical fluctuations in the latter of extremely small amount.

sphere the moon is 27,408 times brighter than *a* Centauri, which is third in brilliancy of all the stars. In our own climate the light of the moon is said to be 3000 times greater than that of the planet Venus.

Planetary light requires fourteen minutes to cross the earth's orbit. According to Bradley, the light of the sun takes eight minutes to reach the earth. Reckoning the mean distance of the sun to be 94,879,956 miles from the earth, it follows that light traverses the air with a velocity of about 200,000 miles per second.* Robins states that a twen-

---

* One hundred and eighty thousand miles per second according to Herschel's calculation, that is 901,000 times faster than sound, which travels at the rate of 1090 feet per second, or nearly thirteen miles in a minute. Were the sun and the earth connected by an iron bar, 1074 days, or nearly *three* years, must elapse before a force applied at the sun could reach the earth. With a bar of tin, nearly *seven* years would be necessary. A writer in the " Quarterly Review " has placed this remarkable property of matter in a popular and paradoxical aspect, by imagining Titan and Saturn placed in opposite points of the orbit of the planet which bears the name of the latter, and their ancient combats being resumed with weapons of earthly fabric, the deadly blow dealt by the former would not slay its victim till after the lapse of fifty-two years; and if, one year before this event, Saturn should aim a mortal thrust at his antagonist, it could not prove fatal till *fifty-one years after his own death!*

ty-four-pounder with a common charge of powder discharges its ball with an initial velocity equal to 1600 feet per second. If such a ball were to continue its velocity undiminished, it would require about *ten* years to traverse a span, which the light of the heavenly bodies pervades in *eight* minutes.

Humboldt, when speaking of this subject, remarks:—"When taking lunar distances from the sun for determinations of geographical longitude, it is not unfrequently found difficult to distinguish the moon's disk among the more intensely illuminated cumuli. On mountains between 13,000 and 17,000 feet high, where in the clearer mountain air only light, feathery, cirrous clouds are to be seen, I found it much easier to distinguish the moon's disk; both being cirrous, from its slighter texture, reflects less of the sun's light, and the moon loses less in passing through the thin atmospheric strata." The ratio of the intensity of the sun's light to that of the full moon deserves further investigation, as Bouguer's generally received determination, $\frac{1}{300,000}$ differs so strikingly from the indeed more improbable one of Wollaston $\frac{1}{800,000}$. Wollaston's comparison of the light of sun and moon, made in 1799, was based on the shadows cast by wax-light, while in the experiments with Sirius, in 1826-27, images reflected from a glass globe were employed. The earlier assigned ratios of the intensity of solar light as compared to that of the moon differ very much from the results here given. Michel and Euler, proceeding from theoretical grounds, have respectively concluded 450,000 and 374,000 to 1. Bouguer, from measurements of the shadows of waxlights, had even made it only 300,000 to 1.

I think, after duly weighing the above facts, we must dismiss from the mind the impression that the intensity of the light of the moon, as compared to that of the sun, has any relation to the supposed morbid effect of lunar light.

I proceed, in the next place, to the consideration of the questions, whether the alleged morbid effect of lunar rays is attributable to something specific in the composition of the light itself; and secondly, whether the supposed abnormal influence of the moon is not altogether owing to certain barometrical variations and meteorological phenomena consequent upon the phases or position of the planet. Is there anything specific in the composition of lunar light?

According to numerous observations which Arago made with his polariscope, the moon's rays contain *polarized* light which *carbonizes*, and is therefore antagonistic to the sun's rays, which *oxygenate*. The polarization of light was not discovered until five years before the death of Huyghens, viz., in 1690. It has, however, been stated that the phenomenon was recognized in 1665, by Grimaldi and Hooke.

When a polarized ray is passed through a thin slice of transparent substance, its effect is said to be most remarkable. "Its whole molecular constitution," says Mr. Hunt, "appears revealed, as if by the touch of a magic wand. The surfaces of many substances, such as thin films of sulphate of lime or mica, became painted with the most beautiful colors. By turning the plate of the polariscope round, these can be made to dissolve or reappear in an exceedingly singular manner. Rings of the most charming colors are seen in some substances, and these are marked with a figure resembling a Maltese cross in black and white.

Polarized light has been used by chemists for the purpose of making subtle analysis of certain articles of food. M. Biot has applied the rays of polarized light to sugar with the view of ascertaining the existence of adulterations. It is easy by means of the polariscope to distinguish between sugar made from beet-root and that manufactured from the sugar-cane.*

What is polarized light? Sir David Brewster thus lucidly explains the phenomenon. When the ray of light falls on a transparent body, so as to be reflected from it, it is modified or affected in such a manner by this reflection, that upon meeting a second transparent body, it will either be reflected or not, according to the side which it presents to it. It will be reflected if it fall upon that body on either of the opposite sides, but will not be reflected if it fall upon either of the other two, at right angles to the former. Thus, suppose the ray, after being modified by the first transparent reflector, presents itself to the second, so as to be reflected, and call the side of the ray, on which it meets the second reflector, on the *north* side; if the second reflector is turned around, so that the *east* side of the ray meets it, there will be no reflection, and in like manner it will be reflected on the south and not on the west sides respectively. The same modification, whatever it may be, prevents the ray from being *doubly* refracted, by passing through Iceland crystal, which it meets on two of its opposite sides, but permits it to be doubly re-

---

* Hunt on Light.

fracted by meeting the crystal on the two other sides. And this modification, with respect to double refraction, may be impressed upon the ray by a first double refraction, as well as by reflection from a transparent body. But where the modification is produced by reflection, it is most complete at one particular angle of incidence, which varies in different transparent substances.

Now, the existence of this phenomenon is certain; it is a fact that a change takes place in the ray by the operation of the first transparent body; it is a fact that this change has some kind of reference to the four sides of the ray, and affects those sides at right-angles to each other differently. The observers of these appearances have explained them, by supposing that each particle of light has its adjacent sides endowed with opposite properties, and that the first reflecting, or double refracting body, turns or arranges all the particles of light in a ray, in such a manner that their similar sides are presented in the same direction to the second body. Now this arranging or turning of the particles, or this change operated by the first body upon the ray, whatever it may be, is termed, from analogy to the phenomena of magnetism—polarization.*

---

* "On Optics," by Sir David Brewster.

Having cursorily referred to two modes of explaining the morbid phenomena of moonlight, I have yet to consider the most rational and philosophic theory of lunar influence propounded —viz., the effect of the moon's position upon the *wind, temperature, and rain*, three meteorological conditions universally admitted to play an important part in the origin, spread, and modification of disease. It has been a vexed question with natural philosophers, whether the barometer is decidedly influenced by the phases of the moon. The facts illustrative of this point are too significant to justify a doubt upon the question.

A remarkable correspondence between the phases of the moon and certain states of the barometer has been observed by Luke Howard. This coincidence, he maintains, consists of a depression of the barometrical line on the approach of the new and full moon, and its elevation on that of the quarters. In above thirty out of fifty lunar weeks in one year, the barometer was found to have changed its general direction once in each week, in such a manner as to be either rising or at its maximum for the week preceding and following about the time of each quarter, and to be either falling, or at its minimum, for two weeks about the new and full. It is remarkable that the point of greatest depression during the year—viz., 28.67, was about twelve hours after the new moon on the 8th of November, and that of the greatest and extraordinary elevation of 30.89, on the 7th of February, at the time of the last quarter. The variation from this coincidence seemed to be owing to an evident perturbation of the atmosphere. These observations were confirmed by observations made for ten years in the Royal Society's apartments. Mr. Howard supposes, therefore, that the joint attractions of the sun and moon at the new moon, and the attraction of the moon predominating over the sun's weaker attraction at the full, tend to depress the barometer by taking off the gravity of the atmosphere, as they produce a high tide in the waters by taking off from their gravity; and again, that the attraction of the moon being diminished by that of the sun at her quarters, this diminution tends to make a high barometer, together with a low tide, by permitting each fluid to press with additional gravity on the earth. It it demonstrated *à priori* on the principles of the Newtonian philosophy, that the air ought to have its tides as well as the ocean, though in a degree as much less perceptible as its gravity.* If this observation were strictly true, and the tides of the atmosphere were to those of the sea as the specific gravity of air is to that of water, the aërial tides must be extremely small, for the weight of air is very trifling compared to that of water. But it is known that the height of the tides of the sea bears some proportion to the extent of the sea, uninterrupted by land, and to its depth. On both these accounts we should expect that the atmosphere would be more influenced by the moon's attraction than the sea, for it is vastly deeper and more extensive than the sea, and entirely unconfined.

Signor Tolado found that a greater elevation of the barometer takes place at the quarters than at the syzygies; it is less when the moon is in the northern signs than in the southern. The mean diurnal height, which corresponds to the Tropic of Cancer, is less by a quarter of a line than that which corresponds to the Tropic of Capricorn. It is one-sixth of a line less at the moon's perigee than at her apogee, and one-tenth of a line less at the syzygies than at the quarters; and there are

---

* "Encyclopædia Britannica."

vacillations in the mercury when the new or full moon corresponds to the apogean or perigean points. He found, also, that the perigee, the new and full moon, and the northern lunistice are favorable to bad weather; while the apogee, the quadratures, and the southern lunistice are more favorable to good weather.

Père Cotte, from observations of thirty-five years, found that the barometer had a tendency to descend at every new and full moon, and to ascend at the quarterly periods. He likewise found that the perigee and northern declination depressed the barometer, whilst the apogee and southern declination had the opposite effect.* Mr. L. Howard has satisfactorily established that the moon's position, operating by the common effects of the attraction of gravitation, influences alike the course of the variable winds, the daily variations of the temperature, and the rain of any year; but not in every year alike, there being a constant periodical variation of the variation itself.†

---

* Orton, on "Epidemic Cholera," p. 222; and "Lectures on Meteorology," by G. Luke Howard.

† M. Kraszewski, of Romanow (as quoted by Andrew Steinmetz in his work on "Sunshine and Showers") has decided, from a long series of observations:—" 1. That 63 times in 100

PART III.

ON THE ALLEGED INFLUENCE OF THE MOON ON THE INSANE.

THE recognized phrase, *luna*tic, is based on the hypothesis that the moon exercises a decided effect upon the insane, and modifies various forms of morbid intellect which do not amount to actual insanity: hence originated the well known legal

---

the weather changes to bad when the moon crosses our equator. 2. When the moon's declination is north, bad weather occurs more frequently than when it is south. 3. When the moon is in its perigee, and at the equator, bad weather occurs 67 times in 100; when in its apogee, but still at the equator, there is bad weather 63 times in 100; and 4. Generally the moon near the equator determines bad weather in the same proportion."

term, "lucid interval."* It was supposed that during paroxysms of mental derangement, the patient was liable to periods of lucidity or mental repose, caused by the various phases of the moon. Acting on this assumption, human life was often made contingent upon a satisfactory solution of the question, was or was there not a lucid interval when a particular crime was committed? The transmission also of property to a vast amount often depended on the reply to a similar interrogatory relating to a period when a certain testamentary disposition was made. A lucid interval legally implies a clear and distinct freedom of the mind from all delusions; in other words, such an intelligent repose and restoration of the intellectual powers as to enable the person mentally disordered to discriminate accurately between right and wrong; thus constituting him morally and

---

* Lunatic, according to ancient legal dicta, is one whose imagination is influenced by the moon and has lucid intervals: " *Lunaticus, qui gaudet lucidis intervallis.*"

Sir W. Blackstone defines a person who is *non compos mentis* to be one " who has had understanding, but by disease, grief, or other accident, has lost the use of his reason; but that a lunatic is indeed properly one who hath lucid intervals, sometimes enjoying his senses, and sometimes not, *and that frequently depending upon the change of the moon.*"—" Commentaries."

legally responsible for his conduct, *quoad* any criminal act of which he may have been guilty, or rendering him competent to exercise a sound memory, judgment, and reflection relating to any testamentary act he has, during a questionable condition of mind, been called upon to execute.

Into the judicial view of this important and vexed question it would be foreign to my purpose to enter.* I therefore propose confining my re-

---

* The rule of law upon this subject is, that all acts done during a lucid interval are to be considered done by a person perfectly capable of contracting, managing, and disposing of his affairs. This has most frequently been a question in wills; and the Ecclesiastical Courts, which first promulgated the rule, adopted it from the Roman law, " *Furiosi autem si per id tempus fecerint testamentum quo furor eorum intermissus est, jure testatum esse videntur.*"—(Justin. Inst., lib. II, tit. 12, s. 1).

A lucid interval consists not in a mere cessation of the violent symptoms of a disorder; neither is it a cooler moment, an abatement of pain or violence, relaxation from a higher state of torture, or the relief of a mind from excessive oppression.

But an interval is that in which the mind, having thrown off the disease, has recovered its general habit. The party must be capable of forming a sound judgment of what he is doing; and his state of mind should be such, that any indifferent person would think him able to manage his own affairs.

M. D'Aguesseau, when pleading before the Parliament of Paris, observed, " That to constitute a lucid interval, there must not be a superficial tranquility or shadow of repose ; but on the con-

marks to the main point at issue, viz., is it satisfactorily established that the moon has any perceptible effect upon the insane?

I should be giving but an imperfect sketch of the literature of this subject if I were not to refer to the fact that the poets, as well as philosophers and medical writers of ancient and modern times, have not failed to countenance by the authority of their genius the popular belief in the influence of the moon not only on the bodily health, but on the passions, and especially on the mind when in a state of disorder. Most of our great dramatic and epic poets have embodied in their immortal

---

trary, a profound tranquility, a real repose; there must be, not a mere ray of reason, which only makes its absence more apparent when it is gone, not a flash of lightning, which pierces through the darkness only to render it more gloomy and dismal, not a glimmering which unites the night to the day; but a perfect light, a lively and continued lustre, a full and entire day, interposed between the two separate nights of the fury which precedes and follows it; and, to use another image, it is not a deceitful and faithless stillness, which follows or forebodes a storm, but a sure, steadfast tranquility for a time, a real calm, a perfect serenity; in fine, without looking for so many metaphors to represent our idea, it must not be a mere diminution, a remission of the complaint, but a kind of temporary cure, an intermission so clearly marked, as in every respect to resemble the restoration of health."

creations this idea. The works of Shakspeare, Spenser, Beaumont and Fletcher, Ben Jonson, Milton, Byron, and Shelley are replete with passages of exquisite beauty in relation to this subject. Our own immortal bard, whose marvellous apprehension and profound knowledge of the mind, intuitive insight into the subtle workings of the human heart and passions, and intimate acquaintance with nearly every branch of knowledge, department of science, art, and philosophy, placed him like a bright and brilliant constellation on a giddy eminence, far above the rest of mankind, has pointedly alluded to the moral and mental influence of the moon on the heart and intellect. In the " Twelfth Night," Viola apostrophizes Olivia as a

"Most radiant, exquisite, and unmatchable beauty."

"I heard you were saucy at my gates," replies Olivia, " and allowed your approach, rather to wonder at you than to hear you. If you be not mad, begone; if you have reason, be brief; *'tis not that time of the moon with me, to make one in so skipping a dialogue.*"

Again, in the play of " Antony and Cleopatra," Enobarbus, after entering Cæsar's camp, thus appeals to the moon :—

> "Be witness to me, O *thou blessed moon!*
> When men revolted shall upon record
> Bear hateful memory, poor Enobarbus did
> Before thy face repent!"

After which, he adds, addressing the moon, previously to expressing his deep contrition for his revolt against Antony—

> "*O sovereign mistress of true melancholy,*
> The poisonous damp of night dispunge upon me;
> That life, a very rebel to my will,
> May hang no longer on me."

In "Othello," after the death of Desdemona, when Emilia enters the chamber to announce the foul murder of Roderigo by the hand of Cassio, the Moor, crushed to the earth by an accumulation of horrible misfortunes, exclaims in the agony of his soul, and in the bitterness of wild despair—

> "It is the *very error of the moon,*
> She comes more near the earth than she was wont,
> *And makes men mad.*"

In "King Richard the Third," the Queen, after rushing, whilst in a state of profound distraction, into the presence of the Duchess of York to announce the death of the King, passionately exclaims—

"Give me no help in lamentation,
I am not barren to bring forth laments;
All springs reduce their currents to mine eyes,
That I, being govern'd by the wat'ry moon,
May send forth plenteous tears to drown the world!"

Milton frequently alludes, in "Paradise Lost," to the morbid effect of the moon. He speaks of

"Demoniac frenzy, moping melancholy.
And *moon-struck madness*."

In Ben Jonson's "Alchemist," Tribulation says—

"But how long time,
Sir, must the saints expect?"

To which Subtle responds—

"Let me see,
How's the moon now? Eight, nine, ten days hence
She will be silver potate; then three days
Before to citronize,—some fifteen days."*

Apart altogether from the astrological ideas prevalent in former times as to the specific influ-

---

* Act iii, scene 1.

ence of the moon on the insane, it is not difficult to understand why this notion should have found favor among learned, scientific, and experienced men. It must be admitted that there is much of the phenomena of insanity yet mysterious and inexplicable. I refer particularly to the remarkable periodicity that in many cases accompanies the progress of this disease. It is commonly supposed to be difficult physiologically and pathologically to explain why such distinct periods of *apparent* recovery should take place during paroxysms of acute insanity. Cases occur in which a patient is outrageously insane one day, and to ordinary observers sane the next; furiously mad one week, calm and composed the next; mischievously deranged in the mind for about a month or six weeks, and then for an equal duration in a sane state of intellect. Not only are these lucid intervals or remissions often noticed during the existence of the mental malady, but the character of the affection also undergoes singular changes and modifications. A patient recently under my care was sad and depressed for one or two days, and then for a similar period was in an exalted and happy state of mental disorder.

Pinel knew a patient who became insane regularly every year. This illness lasted for three months, and terminated favorably towards the middle of the summer. Another patient was annually subject to paroxysms of maniacal furor for a period of fifteen days, but was in the perfect possession of his reason for the remaining eleven months and a half. In another case this physician noticed the attack of insanity to come on every third day, being in this respect analogous to the periodicity observed in tertian intermittent fevers.

In cases of acute melancholia, associated with a suicidal feeling, the attacks often come on periodically. For a week or ten days the desire to commit suicide will prevail, and then the insane craving for self-destruction will suddenly vanish. A patient may be strongly bent on suicide in the morning and have no inclination to destroy himself later in the day.

Drunken madness, or *dipsomania*, as it is termed, in some cases comes on periodically. Bruhl-Cramer cites an instance in which violent paroxysms of intemperance occurred regularly every four weeks at the new moon. Most, another authority, confirms this fact. In Henke's *Zeitschrift für Staatsarzneikunde*, a case is related of monthly periodical attacks of drunkenness, each attack occupying eight days. The patient was ill in this irregular way for seven years. Friedreich says this periodicity is peculiarly characteristic of attacks of dipsomania. In some cases the paroxysms come on with great regularity once in two, three, and four weeks. In other instances several months intervene between the attacks. Occasionally the patient is known to have seizures of dipsomania regularly every second, third, and fourth year. Guislan mentions the case of a woman whose temperate lucid intervals extended over a period of four years. At the end of this time the drunken madness returned, lasting some time with great violence.

Esquirol, Pinel, Falret, Cerise, Brierre de Boismont, Baillarger, Foville, Calmeil, and other celebrated French authorities, cite similar illustrations of the remarkable periodicity so often associated with mental disorders.

Without giving the details of cases that have come under my own personal observation, I may in general terms state that I have frequently observed the phenomenon referred to among insane patients. In this respect, insanity in no way differs from the periodicity which is so characteristic of many forms of general disease; such as intermittent fevers, gout, tic douloreux, cephalalgia, neuralgia, epilepsy, and other affections in which the nervous system is implicated.

Is the periodicity alluded to as often accompanying some types of insanity explicable apart from the hypothesis of lunar influences? I think there can be but one opinion on this subject. The remarkable remissions that occur during attacks of mental derangement depend upon a variety of causes, such as the removal, by appropriate treatment, of temporary congestion of the brain, and irregularity of the cerebral circulation; or the lucid moments may be the effect of paying strict attention to dietetic regimen, condition of the digestive organs, bowels, liver, skin, or kidneys, or be the result of carefully isolating the patient for a time from all mental worry and excitement.

Pinel is of opinion that the periodicity occasionally associated with insanity is more dependent upon an undue indulgence of the angry passions, intemperance in drinking, inanition, or mental agitation caused by a remembrance of the original exciting cause of the malady, than by direct lunar influences. From a general examination Pinel made during the second year of the republic, of the patients confined in the Bicêtre, he found only fifty-two who were subject to paroxysms of insanity at irregular periods, and only six whose attacks were characterized by a regular intermission.

In considering this subject, it is not my intention to quote the somewhat fabulous cases alleged to be illustrative of the moon's influence on the mind, recorded by ancient and modern writers on astrology. Before, however, referring to the opinions of the great Roman authority in medicine, I think the following facts, when viewed in relation to the subject under discussion, worthy of notice: Aristotle records the case of an innkeeper at Tarentum who, although able to attend to his business during the day, became insane soon after the setting of the sun! Bouillon cites the case of a woman who lost the use of her senses at sunset, but who recovered them at daybreak! Sauvage refers to a woman who became maniacal whenever the sun was at its zenith! An ineffectual attempt was made to cure her of the malady by various stratagems, such as keeping her in a dark room and deceiving her as to the hour.

In the chapter treating of Insanity, entitled "*De tribus insaniæ generibus*," Celsus makes no special allusion to the influence of the moon on the insane. In the fourth chapter of his first book, commencing "*De suibus caput infirmum est,*" he says "*Cui caput infirmum est, is, si bene concoxerit, leniter perfricare id mane manibus suis debet; nunquam id, si fieri potest, veste velare; ad cutem tondere: utileque luna.*" He then adds the following advice, by which it is clear that he viewed with grave suspicion the injury likely to arise from an injudicious exposure to the influence of the moon, especially "*before her conjunction with the sun.*" "*Utileque lunam vitare, maximeque ante ipsum lunæ solisque concursum; sed nusquam post cibum.*"

Do the words "*caput infirmum*" mean insanity in the right and generally received acceptation of the term, or do they imply a mere weakness of intellect so often observed after attacks of ordinary fever, and seen to follow other bodily diseases? The latter construction appears to be the one most generally adopted.

The late Dr. Haslam considered it more than probable that the origin of the idea of the moon exercising an influence on the insane may be thus traced and explained: The phases of the moon, and particular female indispositions, are alleged to correspond. The terms used to designate them, have reference to the time when both are completed. Insanity and epilepsy are often connected with these conditions, and suffer an exacerbation at the time when they occur or ought to take place. If, therefore, the ailment referred to in an insane person should be coincident with the full of the moon, and the mind should *then* be more violently disturbed, the recurrence of the same state may be naturally expected at the next full moon. Such has been the prevalence of this opinion, that when patients were brought in former times to Bethlehem Hospital, especially from the country, their friends have generally stated them to be worse at some particular change of the moon, and of the necessity they were under at those times, of having recourse to severe coercion. Some of these patients, after recovering, have stated that the overseer or master of the workhouse himself has frequently been so much under the dominion of this planet (keeping steadily in mind the old maxim, "*venienti occurrite morbo*") that, without waiting for any display of increased turbulence on the part of the lunatics, he has barbarously bound, chained, flogged, and deprived them of food, according as he discovered the moon's age by the almanac!

To ascertain how far this opinion was founded on fact, Dr. Haslam kept, during more than two years, an exact register of the numerous cases under his care, but without finding, in any instance, that the aberrations of the human intellect were influenced by the phases of the moon.

As insane persons, especially those in a furious state, are but little disposed to sleep, even under the most favorable circumstances, they will bestill less so when the moon shines brightly into their apartments.*

Dr. J. B. Woodward, late superintendent of the State Hospital, at Worcester, Massachusetts, at the suggestion of one of the most scientific men in New England, commenced a table of observations on the influence of this planet upon the paroxysms and deaths of the insane, and, after much time devoted to the subject, says :—" These facts and coincidences we leave for the present, with the single remark, that no theory seems to be supported by them, which has existed either among the ignorant or wise men who have been believers in the influence of the moon upon the insane."

The report from which the preceding passage is quoted then adds :—

---

* Haslam's " Observations on Madness and Melancholy," pp. 214—217. London: 1809.

"Many patients are certainly more excitable and restless in pleasant moonlight nights than in dark and gloomy weather; but this would seem to be occasioned by the real or imaginary sight of objects in or without the building, such as men, trees, animals, &c., or the motion, perhaps, of the passing clouds. An opinion, however, that has existed for so long a period, which has spread so extensively, and which, in this country, is familiar as 'household words,' deserves to be carefully examined; for it is important to disprove error as well as to establish truth.

"Weak and timid females are sometimes alarmed and much agitated during the continuance of lightning and thunder; but as a general thing, we have not observed the insane to be much disturbed on such occasions."

In the preceding passages I have endeavored accurately to record the opinions of various accredited medical authorities adverse to the theory of the moon's influence upon the insane.

"*Ogni medaglio ha il suo reverso,*" says the well-known Italian proverb. So it is with the vexed question under consideration. I therefore proceed to an analytical examination of the conclusions which men of great experience, of undoubted eminence and veracity, have arrived at in favor of an opposite hypothesis.

In the foremost rank among those who advocate the lunar theory, the illustrious Frenchman, Pinel, justly occupies a proud, elevated, and conspicuous position. The scientific world bows with profound respect and reverence to every opinion to which his masterly intellect gives expression. This eminent physician thus addressed himself to the subject: "It is curious to trace the effects of sollunar influence upon the return and progress of maniacal paroxysms. They generally begin immediately after the summer solstice, are continued with more or less violence during the heat of summer, and commonly terminate towards the decline of autumn. This duration is limited within the space of three, four, or five months, according to deficiency of individual sensibility, and according as the season may happen to be earlier, later, or unsettled as to its temperature. Maniacs of all descriptions are subject to a kind of effervescence or tumultuous agitation upon the approach of stormy or very warm weather. They then walk with a firm but precipitate step; they declaim without order or connexion; their anger is roused by trivial or imaginary causes, and they express their feelings by clamorous and intemperate vociferation.* We must not extend this law of solar influence beyond its natural boundary, nor conclude that the return of maniacal paroxysms is invariably dependent upon the atmosphere. I have seen cases in which the paroxysms return upon the approach of winter, *i.e.*, when the cold weather of December and January set in; and this remission and exacerbation corresponded with changes of the temperature of the atmosphere from mildness to severe cold."

The preceding quotation from Pinel's treatise on insanity is entitled to profound respect. His sphere of observation was great, his experience large, his veracity unimpeachable, and what he saw he truthfully recorded. I think, therefore, we are bound, even in opposition to our own judgment,

---

* In the year 1807, Mr. Thornton, then one of the apothecaries of Bethlehem Hospital, paid particular attention to the influence of the morning light upon all the patients that were at the time confined in the asylum. He came to the conclusion that many of them became noisy as soon as the day began to break, and that, with the exception of two or three recent cases, they all became silent and quiet after night. During the eclipse of the sun on the 16th of June, 1806, there was a sudden and total silence in all the cells of the hospital!

to bow with great deference to the conclusion at which this distinguished physician arrived with regard to the influence of the moon on the insane.

Daquin, another eminent physician, directed his attention to this subject. He occupied a high professional position in France, and was universally respected for the zeal with which he cultivated science, and the faithful accuracy displayed in recording the results of his experience. He was a close observer of nature, a man of truth, and of highly-cultivated understanding.

In 1791 he published an able work on insanity. In this he enters at length into a discussion of this vexed question. Without quoting in detail the remarks of Daquin, I shall content myself with generalizing the data brought forward in support of the opinion he has formed with reference to the influence of the moon upon the mind when in a state of aberration. He says:—

" It is a well-established fact that insanity is a disease of the mind upon which the moon exercises an unquestionable influence. The new moons and the last quarters of the moon are the lunar phases which influence the insane most frequently and painfully.

" The first quarters and the full moons are the

phases which I have observed to have the least influence in inducing relapses of insanity; the insane at these periods being less insane and quieter, and they reasoned almost as if they were not ill at all.

"Those who are still susceptible of being cured, as well as those who have been cured, are precisely those upon whom the two most powerful lunar phases have had the greatest influence during the whole of their illness.

"Those who are acutely maniacal are much more susceptible to the influence of the lunar phases than others.

"I have also observed a difference between the influence exerted by this planet on madness characterized by excessive joy, and that by sorrow and melancholy.

"It is proved that this influence is much more marked in parts of the countries bordering on the sea than in those at a distance from it.*

Amidst this conflict of testimony, what conclusions can legitimately be drawn? Are we to consider the theory of lunar influence as a myth and an idle fable, or as a well-established fact based

---

* Daquin, " Philosophie de la Folie," 1st edition, 1791.

upon accurate and scientific data? It is impossible altogether to ignore the evidence of such men as Pinel, Daquin, Guislain, and others, and yet the experience of modern psychological physicians is to a great degree opposed to the deductions of these eminent men.* Is it not probable that there is some degree of truth on both sides of the question; in other words, that the alleged changes observed among the insane at certain phases of the moon may arise, not from the direct, but the indirect influence of this planet? It is well known that certain important and easily recognizable meteorological phenomena result from the varied positions of the moon; that the rarity of the air, the electric conditions of the atmosphere, its degree of heat, dryness, moisture, and amount of

---

* Guislain has recorded the history of an insane patient on whom the influence of the moon was observed. He became maniacal every twenty-eight days. This distinguished physician says, " We have among our female patients a maniac sixty years of age. Her attacks of acute insanity are periodical; the return of the disease corresponding with the return of the full moon."
—[*Leçons Orales sur les Phrénopathies*, Gand, 1852.]

In a work entitled, " De l'Électricité du Corps Humain," by Mons. l'Abbé Bertholon, the case of a lunatic is recorded whose periodical accessions of acute mania occurred invariably at the full of the moon.

wind prevailing, are all, more or less, modified by the state of the moon. In the generality of bodily diseases what obvious changes are observed to accompany the meteorological conditions referred to? Surely those suffering from diseases of the brain and nervous system affecting the mind cannot, with any show of reason, be considered as exempt from the operation of agencies that are universally admitted to seriously affect patients afflicted with other maladies? That the insane do appear to a degree unusually agitated at the full of the moon, particularly if its bright light is permitted uninterruptedly to enter the room where they sleep, there cannot be a doubt.* This phenomenon may, I think, be accounted for apart altogether from the hypothesis of there being anything *specific* in the composition of the lunar ray.

In certain forms of insanity, particularly those characterized by illusions of the senses or hallucinations of the mind, how materially affected

---

* An intelligent lady, who occupied for about five years the position of matron in my establishment for insane ladies, has remarked, that she invariably observed a great agitation among the patients when the moon was at its full. This must be accepted *quantum valeat.*

the patients are by the kind and degree of light admitted into their chambers. If exposed to a great degree of natural, or even artificial light, the hallucinations often become painfully intensified. Sleeplessness often thus arises, and dormant morbid visual and aural conditions previously in a latent state, become actively developed. Patients afflicted with distressing delusions often imagine that everything they see has assumed the form of a terrible spectral image. The presence of light emanating from the moon, when at its full, greatly tends in these cases to aggravate the lunatic's mental sufferings. It is occasionally most desirable to exclude light altogether from bedrooms occupied by the insane, in order to tranquilize them and cause sleep. In other instances the patient's condition of mental excitement becomes aggravated by being kept for a length of time in darkness. It is therefore often necessary, with the view of dissipating a false creation which has seized hold of his disordered imagination, to admit light freely into the chamber. This has been known immediately to compose the patient, by convincing him of the morbid and illusory character of his mental impression, and inducing him to exclaim,—

> "There is no such thing!
> It is [*my disordered brain*]* which informs
> Thus to mine eyes."†

If I were to confine the concluding remarks to the result of my own observations on this subject, I fear I should be able to add but little to what others have written in relation to it. I freely admit that, placing but little faith in what has been recorded or said on the subject, I have not kept any systematic register as to the effect of different phases of the moon on the insane.

It has been observed by some officials connected with lunatic asylums, that occasionally, from causes unknown to them, many of the patients have simultaneously exhibited a state of unnatural agitation. This condition of the inmates has occurred several times during the year, and at irregular intervals; but the state of the moon at the time was not made a matter of observation. It is difficult to assign any cause for the alleged phenomenon,

---

* I hope I shall not expose myself to critical censure for substituting the words "disordered brain" for "bloody business."

† Shakspeare has with exquisite poetic beauty and wonderful psychological truth delineated, in the passage from which the preceding lines are taken, the sudden transition of the mind from apparent insanity (assuming the form of an illusion of the senses) to that of health.

and it would not be entitled to a moment's consideration if the commotion and excitement that are said to take place were confined to a few of the patients; but as many of the insane inmates (several of them occupying separate rooms) were noticed to be thus affected, at the same day and hour, the fact becomes one worthy of record.*

* There can be no doubt that *bright red*, and *yellow* rays stimulate and in some cases irritate the brain and mind. *Deep blue* is said to depress or exhaust the vital force. Some animals are excited when brought in contact with *scarlet* color. This is observed at the bull-fights that take place in different parts of Spain. *Green, violet,* and in fact all the neutral tints, soothe the nervous system and allay mental irritation. Those who have the care of the insane, or medical treatment of patients suffering from great brain irritability, should bear these facts in remembrance.

PART IV.

HYGIENE OF LIGHT.

HAVING in the introductory part of this work, referred in general terms to the sanitary influence of light, I return to the subject in order to point out more specifically the deleterious effect of its absence on the bodily health. I have previously alluded to the important changes that take place in the constitution of the blood, in consequence of the cutaneous vessels on the surface of the body not being freely exposed to the oxygenating and life-generating influence of the sun.

It is a well-established fact that, as the effect of isolation from the stimulus of light, the fibrine, albumen, and red blood-cells become diminished in quantity, and the serum, or watery portion of the vital fluid, augmented in volume, thus inducing a disease known to physicians and pathologists by the name of *lukæmia*, an affection in which *white* instead of red blood-cells are developed.* This exclusion from the sun produces the sickly, flabby, pale, anæmic condition of the face, or exsanguined ghost-like forms so often seen among those not freely exposed to air and light. The absence of these essential elements of health deteriorates by materially altering the physical composition of the blood, thus seriously prostrating the vital strength, enfeebling the nervous energy, and ultimately inducing organic changes in the structure of the heart, brain, and muscular tissue.* Do not these facts suggest the great importance, particularly in a northern climate like that of England, where the

---

* Virchow's "Cellular Pathology," 1860.

* "Where light is not permitted to enter the physician will have to go," is the translation of a well-known Italian proverb. In the eloquent words of Sir David Brewster, " Light is the very life-blood of nature, without which everything material would fade and perish ; the fountain of all our knowledge of the external universe ; and the historiographer of the visible creation, recording and transmitting to future ages all that is beautiful and sublime in organic and inorganic nature, and stamping on perennial tablets the hallowed scenes of domestic life, the ever-varying phases of social intercourse, and the more exciting scenes of bloodshed and war, which Christians still struggle to reconcile with the obligations of their faith."

bright beams of the sun so seldom shine, of so constructing our habitations, both with regard to the number of windows and position of the buildings, as to admit within their walls a sufficient degree of air and light? and are we not bound to impress upon the legislature the necessity of enacting some stringent protecting measures, with a view of preventing low damp cellars and rooms being crowded together, into which air and light vainly struggle to penetrate, and where so many thousands of the poorer classes in our large towns are compelled to dwell, to the utter sacrifice of every comfort worth living for, and positive ruin to the health of both body and mind?* In several Eng-

---

* The following is an extract from a recently published report of the health commissioners of the State of New York: "In no city in the civilized world is there to be found half a million of people so unhealthily housed, as a class, as the tenement-house population of New York. In less than sixteen thousand houses, on lots that average scarcely twenty-five by one hundred feet, there dwell nearly five hundred thousand people; and in the cellars of those houses nearly sixteen thousand more poor, whose poverty and ignorance allow no election of better homes."

The Rev. Isaac Taylor, incumbent of St. Matthias, Bethnal Green, thus describes in a printed circular the terrible state of the poor of his parish. The statement is copied from the *Pall Mall Gazette* of February 12: "The mortality among young children is something frightful. I do not know anything more

lish manufacturing cities, until within the last few years, the great proportion of the working classes lived either under-ground, in courts, or in narrow streets, and houses almost destitute of windows, apparently constructed for the specific purpose of admitting the *minimum* portion of air and the smallest possible degree of light consistent with the preservation of health and the maintenance of life.*

terrible than the statements which one continually hears. It is a common thing for a mother to say that she has buried six or eight, and reared one or two. This mortality among the children is chiefly owing to the deadly over-crowding, and to insufficiency of food and clothing. Last summer we found a family of eight children living with their father and mother in a room some ten feet square, and almost in a state of starvation. All the children had the small-pox out upon them; they had no medical care or nursing; the only medicament that had been used was a little oil rubbed over their faces; this the father said he had heard was good for the small-pox. The man was engaged meanwhile in the delicate work of making white chenille, to be sold in the fashionable West-end shops. Hardly a family in the parish possess more than a single room, in which all the members live, work and sleep."

* Sir David Brewster, in his opening address delivered in November of last year before the *Royal Society of Edinburgh*, has suggested an easy and valuable remedy for these frightful evils. His remarks afford a valuable illustration of the facility with which the great truths of science may be practically applied in promoting the comfort and health of the human race: " If, in a very narrow street or lane, we look out of a window with the eye in

It was computed not many years ago that in Liverpool between 30,000 and 40,000 people lived in cellars. As a consequence of this state of things, the health of the working classes became seriously affected. Legislative measures were adopted for the purpose of declaring such habitations illegal, and those living in them were ejected by the strong arm of the law. In 1849, 4,700 cellars were cleared of 20,000 inhabitants.

Rightly recognizing the life-giving and life-sustaining influence of light, the " Towns Improve-

---

the same plane as the outer face of the wall in which the window is placed, we shall see the whole of the sky by which the apartment can be illuminated. If we now withdraw the eye inward, we shall gradually lose sight of the sky till it wholly disappears, which may take place when the eye is only 6 o: 8 inches from its first position. In such a case the apartment is illuminated only by the light reflected from the opposite wall, or the sides of the stones which form the window; because, if the glass of the window is 6 or 8 inches within the wall, as it generally is, not a ray of light can fall upon it. If we now remove our window, and substitute another in which all the panes of glass are roughly ground on the outside, and flush with the outer wall, the light from the whole of the visible sky, and from the remotest part of the opposite wall, will be introduced into the apartment, reflected from the innumerable faces or facets which the rough grinding of the glass has produced. The whole window will appear as if the sky were beyond it, and from every point of this luminous surface light will radiate into all parts of the room."

ment Clauses Act of Scotland" enacted that no cellars less than seven feet high, without a window, and of which more than two-thirds are below the level of the street, should be inhabited. Acting upon the authority of this statute, the corporation of the city of Edinburgh has considerably improved the physical condition of the working classes of that town by peremptorily closing 3,000 dreary, sunless dens, in which the poor people were in the habit of residing.

How eloquently and truthfully has Sir David Brewster illustrated this section of my subject. He says: "If the light of day contributes to the development of the human form, and lends its aid to art and nature in the cure of disease, it becomes a personal and national duty to construct our dwelling-houses, schools, workshops, factories, churches, villages, towns, and cities upon such principles and in such styles of architecture as will allow the life-giving element to have the fullest and the freest entrance, and to chase from every crypt, cell, and corner the elements of uncleanness and corruption which have a vested interest in darkness.

"Although I have not visited the prisons and lazarettos of foreign countries, to describe the dungeons and caverns in which the victims of despotism and crime are perishing without light and air, yet we have seen enough in our own country—in private houses, in the most magnificent of our castles, and in the most gorgeous of our palaces—to establish the fact that there is hardly a house in town or country without dark apartments which it is in the power of science to illuminate. In most of the principal cities of Europe, and in many of the finest towns of Italy, where external nature wears her brightest attire, there are streets and lanes in which the houses on one side are so near those on the other, that hundreds of thousands of human beings are neither supplied with light nor air, and carry on their trades in almost total darkness. Providence—more beneficent than man—has provided the means of lighting up to a certain extent the workman's home, by the expanding power of the pupil of his eye, and by an increasing sensibility of his retina; but the very exercise of such powers is painful, and every attempt to see when seeing is an effort, or to read and work with a straining eye and an erring hand, is injurious to the organ of vision, and sooner or later must impair its powers. Thus, deprived of the light of day, thousands are compelled to carry on their trades principally by artificial light—by the consumption of tallow, oil, or gas—thus inhaling from morning till midnight the offensive odors and polluted effluvia which are more or less the products of artificial illumination.

"It is in vain to expect that such evils, shortening and rendering miserable the life of man, can be removed by legislation or arbitrary power. In various great cities attempts are making to replace their densely congregated streets and dwellings by structures at once ornamental and salutary; and Europe is now admiring that great renovation in a neighboring metropolis, by which hundreds of streets and thousands of dwellings, once the seat of poverty and crime, are replaced by architectural combinations the most beautiful, and by hotels and palaces which vie with the finest edifices of Greek or of Roman art."*

Foucault cites a striking illustration of the sad effects of the absence of light on the health of young children. On one occasion his attention was attracted to the mutilated condition of several large mulberry trees, the branches of which, before their decay, effectually shaded the schoolroom in

---

* "An Address delivered at the Opening of the Session for 1866-7, at the Royal Society, Edinburgh."

which a number of orphan girls affected with chronic diseases were educated. On asking the reason for the mutilation, he was informed that the shade of the trees visibly increased the severity of the scorbutic affections that prevailed among the children, and that a very favorable change had taken place in the condition of the girls since their exposure to the unimpeded light of the sun.*

Utterly regardless of the principles just enunciated, how often do we see parents, who cannot for a moment be considered indifferent as to the present and future health of their offspring, adopting the most ingenious means of effectually excluding light from the bodies of young infants and children! No course can be more detrimental to their health than the one just referred to, because the value of an important vital element is systematically ignored at a period of life when it is of the highest importance it should be brought to bear upon the purification of the blood, and consequent healthy development of organic structures.

Children, even at an early age, should not be excluded, particularly during the warm periods of

---

* "Causes Générales des Maladies Chroniques," p. 42.

the year, from the genial and cheering influence of the sun. The sanitary effect of the light can be easily made available even during the winter months (in rooms properly ventilated and heated) with little or no danger. Great benefit to the health would accrue by giving children what the ancients termed *solaria*, or " solar air-baths ;" that is, permitting them to lie naked upon the bed or floor, free from the incumbrance of swaddling-clothes, so that their bodies may be thoroughly brought under the influence, for some period of the day, of good air and bright sunlight. The children of savages, as well as of negroes, who are often allowed, as soon as they can walk alone, to run about in the open air (*in puris naturalibus*) freely exposed to the influence of light, have finely-developed muscular structures, and generally enjoy robust health.*

---

* " Passing round St. Paul's Churchyard the other day, our ear caught a low confused sound, which seemed to issue from behind a *grille* under the gloomy portico of St. Paul's School. Attracted by so unusual a sound, and blinded for a moment by passing from the sunlight under the portico, we at length made out that the sound was that of a boy's voice in play. After a little time, when our eyes had become accustomed to the darkness, and permitted us to penetrate the dreary recess, we perceived that some score of boys were actually at play within its unhappy shade. Such a

No amount of artificial, polarized, or reflected light will compensate for the want of the direct action of the sun.

Humboldt, in the account he has published of his voyage to the equinoctial regions, says, when speaking of the Chaymas, " Both men and women (whose bodies are constantly inured to the effect of light) are very muscular, and possess fleshy and rounded forms. It is needless to add that I have not seen among these people a single case of natural deformity. I can say the same of many thou-

---

profound sanitary error on the part of the authorities of the school as to shut up their youth in a cage, without light or air, and there bid them play, struck us with some astonishment, Boys at their games are like young birds singing on the bough ; half their joy is a result of the influence of the sunny air. We are told, indeed, that only a very short time is spent in this gloomy retreat ; but surely it is hard that the hour of recreation, when academic art should give way to nature, should be selected for this depressing process as at present carried on. What would Dean Colet have said to it? When the St. Martin's national school, leading out of Endell street, was built some years ago, we noticed with pleasure that a playground was built at the top of the school, where light and air were plentiful, conditions essential in a playground where children use violent exercise ; but this example we are sorry to see has no effect upon the Mercers' Company. The necessity of light for young children is not half appreciated. Many of the affections of children, and nearly all

sands of Caribs, Maysias, Mexicans, and Peruvian Indians, whom I observed during five years. Deformities and deviations from healthy physical development are exceedingly rare in certain races of men, especially those which have the skin strongly colored, who wander about naked under the brilliant light of the tropical regions. These have muscular fleshy bodies, rounded contours, and present none of those deformities so frequently observed among the inhabitants of other climates."

Dr. Bryson, in his memorandum to the Lords

---

the cadaverous looks of those brought up in great cities, are ascribable to this deficiency of light and air. When we see the glass-rooms of the photographers in every street high up on the topmost story, we grudge them their application to a mere personal vanity. Why should not our nurseries be constructed in the same manner? If mothers knew the value of light to the skin in childhood, especially to children of a scrofulous tendency, we should have plenty of these glass-house nurseries, where children may run about in a proper temperature, free of much of that clothing which at present seals up the skin—that great supplementary lung—to sunlight and oxygen. Glass-house nurseries lifted up to the topmost story would save many a weakly child that now perishes for want of those necessaries of infant life."

I copy the preceding sensible remarks, so strictly in conformity with my own views, from the *Pall Mall Gazette*. They are from the pen of Dr. Andrew Winter.

Commissioners of the Admiralty relating to the health of seamen, has pointed out the great difference between the appearance of sailors employed in the dark bread-room and hold of ships, and those who are freely exposed to air and sunlight on deck, or in open boats, at all hours of the day; he therefore considers it of importance to "ascertain whether exclusion from the solar rays be not, to a greater extent than is generally believed, one reason why these men, who have in consequence acquired a pale waxy look from confinement below, are more susceptible to disease, and less capable of sustaining its shocks, than are those whose blood is enriched and strengthened by the free exposure to light, heat, and air which their different avocations ensure. The force of these remarks, however, will be best understood by those who have had opportunities of witnessing the rapid change which takes place in the human constitution by exposure for only a short time to the direct rays of a tropical sun. Why, in a state of perfect repose, the blood should acquire a brighter tinge and an increased force of circulation, are inquiries the value of which the observant physiologist will not fail justly to appreciate; neither will he fail, as often as opportunities occur, to follow up these phenomena, should they terminate in disease, or unhappily produce death."*

The inestimable value of light as an element in the preservation of health and treatment of disease, should be fully appreciated in the construction of all streets and buildings, particularly those intended as habitations for the poor, or public hospitals for the treatment of disease. It is a well-ascertained fact that many maladies are more susceptible of amelioration, if not of cure, provided the light of the sun is freely admitted into the rooms or wards where invalids are domiciled.*

Apart altogether from the cheerfulness and mental serenity (important auxiliaries in the eradication

---

* " Manual of Scientific Inquiry," published by authority of the Lords Commissioners of the Admiralty, edited by Sir John F. W. Herschel, Bart. London, 1849.

* Some years ago, when visiting the great hospital of St. John, at Brussels, I was much impressed with the most judicious hygienic arrangements made in the construction of that building for the health and comfort of the invalids. On the roof of the hospital an elegant garden is laid out with great taste, and planted with shrubs, small trees, and a grass lawn, interspersed with pretty and sweet-smelling flowers. It is, in fact, a *Parc de Monceau* in miniature. In this quiet rural retreat, patients, particularly the convalescents, are permitted at certain hours of the day to promenade, indulging in the luxury of good air and bright sunlight.

of disease!) which the bright rays of the sun invariably engender, light has a thermic influence upon the mind and body when prostrated by serious ailments, and certainly acts beneficially by chemically purifying the blood of the patient, as well as the atmosphere of the apartment he occupies.

There are, of course, active conditions of bodily and mental disease, such as small-pox, inflammatory conditions of the skin and brain, acute mania, ophthalmia, etc., which require to be carefully excluded from the stimulus of light. I have elsewhere addressed myself to this subject.

Florence Nightingale has entered so fully into this subject, that I offer no apology for quoting *in extenso* her extremely judicious remarks relative to the importance of keeping prominently in view, in the architectural construction of public hospitals, the sanitary value of light to the sick. She says:

"Second only to fresh air, however, I should be inclined to rank light in importance for the sick. Direct sunlight, not only daylight, is necessary for speedy recovery; except, perhaps, in certain ophthalmic and a small number of other cases. Instances could be given, almost endless, where, in dark wards or in wards with a northern aspect, even when thoroughly warmed, or in wards with borrowed light, even when thoroughly ventilated, the sick could not by any means be made speedily to recover.

"Among kindred effects of light I may mention, from experience, as quite perceptible in promoting recovery, the being able to see out of a window, instead of looking against a dead wall, the bright colors of flowers; the being able to read in bed by the light of a window close to the bed-head. It is generally said that the effect is upon the mind. Perhaps so; but it is no less so upon the body on that account.

"All hospital buildings in this climate should be erected so that as great a surface as possible should receive direct sunlight—a rule which has been observed in several of our best hospitals, but, I am sorry to say, passed over in some of those most recently constructed. Window-blinds can always moderate the light of a light ward; but the gloom of a dark ward is irremediable.

"The axis of a ward should be, as nearly as possible, north and south; the windows on both sides, so that the sun shall shine in—from the time he rises till the time he sets—at one side or the other. There should be a window to at least every two beds, as is the case now in our best hospitals.

Some foreign hospitals, in countries where the light is far more intense than in England, give one window to every bed. The window-space should be one-third of the wall-space. The windows should reach from two or three feet from the floor to one foot from the ceiling. The escape of heat may be diminished by plate or double glass. But while we *can* generate warmth, we cannot generate daylight or the purifying and curative effect of the sun's rays."*

Many facts are on record, illustrative of the decidedly beneficial effect of the free admission of light into public buildings set apart as receptacles for cases of disease.

Sir James Wylie, of the Imperial Russian Service, pointed out to an English physician one of the barracks at St. Petersburg, in which three cases of disease occurred on the dark or shaded side of the establishment for one on the other, though the apartments on both of these sides communicated freely with each other, and the discipline, diet, and treatment were in every respect the same.*

---

* "Notes on Hospitals," by Florence Nightingale. Third edition, pp. 19, 20. London: Longmans, 1863.

* "On the Theory and Practice of Ventilation," by R. B. Reid, M. D. 1855.

A very remarkable instance of recovery from disease has been related by the late Baron Dupuytren, the eminent French surgeon, and cited in a work on Light published by the Society for Promoting Christian Knowledge. A lady residing in Paris had suffered for many years from an enormous complication of diseases, which had baffled the skill of all her medical advisers, and her state appeared almost hopeless. As a last resource, the opinion of Dupuytren was requested upon her case, and he, unable to offer any direct medical treatment essentially differing from all that had been previously tried in vain, suggested that she should be taken out of the dark room in which she lived, and away from the dismal street, to a brighter part of the city, and that she should expose herself as much as possible to the daylight. The result was quickly manifest in her rapid improvement, and this continued until her recovery was complete. An equally singular instance has been related by Southey, in the case of his own parent.*

---

* " In chlorosis, scrofula, phthisis, and in general, every disease characterized by deficiency of vital power, light should not be debarred the patient. In convalescence from almost all diseases it acts, unless too intense or too long continued, as a

"In the years of cholera, when this frightful disease nearly decimated the population of some of the principal cities in the world, it was invariably found that the deaths were more numerous in narrow streets and northern exposures, where the salutary beams of light and actinism had seldom shed their beneficial influence. This resistless epidemic found an easy prey among a people whose physical organization had not been matured under those benign influences of solar radiation which shed health and happiness over our fertile plains, our open valleys, and those mountain sides and elevated plateaus where man breathes in the brighter regions of the atmosphere." *

I have spoken previously of the inestimable benefits which accompany a liberal exposure of the

---

most healthful stimulant, both to the nervous and physical systems. The evil effects of keeping such invalids in obscurity are frequently very decidedly shown, and cannot be too carefully guarded against by the physician. The delirium and weakness which are by no means seldom met with in convalescents kept in darkness, disappear like magic when the rays of the sun are allowed to enter the chamber. I think I have noticed that wounds heal with greater rapidity when the light is allowed to reach them than when they are kept continually covered."—(Dr. Hammond's "Treatise on Hygiene.")

* Sir D. Brewster.

body to the action of the solar beam. I cannot pass unnoticed the positive mischief that often arises from a prolonged action of intense light on the visual organs, as well as the injury often caused by a sudden transition from a moderate to a great or glaring degree of the sun's influence.*

Soldiers are said often to suffer in their eyes from the reflection of the rays of the sun from

---

* A sudden transition from perfect darkness (particularly if ot considerable duration) to an intense degree of light, produces a most painful and dazzling sensation in the retina, accompanied by a feeling of acute pain in the eyeballs, as well as of the head. The symptoms are often associated with temporary blindness. The tyrant Dionysius, being aware of these facts, inflicted on some of his unhappy prisoners the severe punishment of compelling them to pass rapidly from great darkness to the full blaze of the sun in recently whitewashed rooms!

An officer of high rank in the army of Charles I. went to Madrid, with the view of executing an important mission for the king, but failing in his generous attempt to do his Majesty a signal service, he was arrested by the Spanish Government, and ordered to be incarcerated in a dark and dismal dungeon, into which light never entered except when a small hole at the top of the cell was opened by the jailer for the purpose of giving the prisoner food. This unfortunate royalist was confined in the dungeon for several months: he was then set at liberty. Such, however was the effect of the darkness upon his nervous system, eyes, and optic nerve, that he was compelled to live for some time afterwards in a dark room, in order gradually to accustom himself to the stimulus of light.

white sand or snow. Lévy records that in 1819 the Swiss troops in garrison at Lyons had many of their number affected with hemeralopia, accompanied with nervous symptoms, such as nausea and vomiting, caused, it is supposed, by drilling under a hot sun.*

This disorder of the vision is said to arise from two very different causes: 1st, an exposure to a stronger light than the eye has been accustomed to; 2d, a deficiency of the black pigment which lines the choroid membrane. This affection of the eye is commonly observed among those who live in dark caverns, mines, dungeons, or prisons. It is frequently seen among the peasants of Italy, who are employed in agricultural pursuits. This is attributed to the peculiar brightness of the Italian sky, remarkable clearness of the atmosphere, and relaxingly warm condition of its temperature. The peasants of Italy are thus exposed to the joint operation of almost every cause that can produce habitual debility of the iris and irritability of the retina. This complaint is most frequent on return of spring, or at the vernal equinox.

---

* A form of intermittent blindness, characterized by the patient being able to see only in broad daylight, and becoming totally blind at sunset.

Ramazzini says that such is the degree of dimness of vision which this exposure to intense light induces, that the peasants lose their way in the fields in the glare of noon, but that on the approach of night they are again able to see distinctly. In the treatment of these cases it is found necessary to keep the patients for some weeks in the shade, or in comparative darkness, until the eyes recover their healthy tone. This disease of the eye is said to be endemic in some parts of France, particularly in the neighborhood of Roche Guion, on the banks of the Seine. So generally does it prevail, that we are told upon good authority that it affects one in twenty of the inhabitants in one village, and in another, one in ten every year. It shows itself in the spring, and continues for three months, returning in a slighter degree in the autumn. The soil of this part of France consists of dazzling chalk, and it is thought that the intensity of the first reflected light after the dreariness of the winter in all probability is the cause of the malady.

This is a common affection in Russia during the summer months, when the eyes are exposed almost without intermission to the constant action of light, as the sun then dips but little below the horizon, and there is scarcely any interval of darkness. The peasants who protract their hard labor in the fields from a very early to a very late hour, and at the same time exhaust and weaken themselves by their daily fatigue, are subject to this malady. Mr. Guthrie has published an account of a detachment of Russian soldiers, in which this disorder suddenly manifested itself. They were ordered to attack a Swedish post at the moment the affection of the eyes developed itself, and had nearly destroyed one another by mistake, owing to their visual powers being impaired. These soldiers had been harrassed by long marches, and exposed night and day to the piercing glare of an uninterrupted range of snowy mountains.*

"The only persons I have myself seen affected with hemeralopia have been those just returned from sea voyages—most commonly from the East or West Indies—and who have consequently been exposed to a strong glare of sunlight. The affection is, I believe, also met with among the inhabitants of the inland parts of India, who attribute

---

* " Mém. de la Société Royale de Méd., 1786." " Memoirs of the Medical Society of London," by Mr. Guthrie. " The Study of Medicine," by John Mason Good, M.D., F.R.S.

it, just as our own sailors do, to sleeping when exposed to the moonbeams.

"The real cause of hemeralopia appears to be exhaustion of the nervous susceptibility of the retina from over-excitement by the sun's rays, whereby the part is rendered incapable of appreciating the milder rays of twilight or moonlight.

"But this exposure to strong light is not always the cause of the affection; for I have met with it among those who had never quitted the temperate parts of the globe. In most of the latter cases, however, the complaint has shown itself after voyages which had subjected the patients to exhausting labor, and exposure to severe weather, when deprived of their proper supply of fresh provisions and vegetables.

"I have commonly found that a few weeks' residence on shore, with a wholesome *mixed* diet and the use of quinine, has restored their vision to a healthy state.*

This disorder of the vision is generally associated with a scorbutic condition of the system. It prevailed to a considerable extent among the

---

* "A Guide to the Practical Study of Diseases of the Eye," by James Dixon, F.R.C.S., England, Surgeon to the Royal London Ophthalmic Hospital, Moorfields, London, 1866.

French and English soldiers during the Crimean campaign. It no doubt arose from their long-continued exposure to an intense glare of light, which their eyes had not been accustomed to. An eminent American ophthalmic practitioner says that a surgeon attached to the garrison of Strasbourg, in France, has proposed what he claims to be a speedy and effectual mode of curing this malady.* He confines his patients in an entirely dark room for a considerable number of hours, not allowing light to enter even for an instant. The eyes thus being left in complete repose, recover their normal tone. Relapses, he asserts, are quite rare, provided the seclusion has been of sufficient duration. Soldiers suffering from this ailment are advised to be removed into the interior of the country from seaboard stations, particularly if exposed there to the glare from white sand.

Snow and ice blindness is a common affection among those who reside for a considerable time in Alpine or Arctic regions.†

Sheep are known to temporarily lose their sight

---

* Dr. Henry W. Williams.

† Dr. Hammond knew a child who was rendered permanently blind by looking intently at a bright object while she was having her photograph taken.

if permitted to wander over mountains covered with snow; but it returns a short time after the snow and ice are melted.

The Greek soldiers, as related by Xenophon, suffered severely in their eyes from the reflection of the intense light of the sun as they crossed the mountains of Armenia.

It is impossible, whilst considering this matter, to leave altogether unnoticed the hygienic treatment of the eyes, when morbidly affected by artificial light.

Persons exposed for an undue length of time to the glare of brilliantly-lighted rooms, often suffer from chronic ophthalmia and other affections of the organ of vision. Literary men, from the same cause, are liable to attacks of *muscæ volitantes* and *amaurosis*. Tailors, sempstresses, shoemakers, jewellers, watchmakers, and, in fact, all who work by artificial light, are subject to serious disorders of the eye. Under these circumstances they often become acutely sensitive to light.* According to

---

* Mr. Ernest Hart says, that when great sensitiveness to light exists, tinted (plain) spectacles are indicated. He observes: " The ordinary smoked glasses may be advantageously substituted in such cases by those tinted of a cobalt blue, which sufficiently exclude the irritating yellow and red rays, without cutting off too

a distinguished oculist, light is injurious to the eyes in proportion as the *red* and *yellow* rays prevail. These produce cerebral and visual excitement, followed by debility of the retina.* He suggests as a remedy for the injurious effects of *red* and *yellow* colors that the light should be surrounded by a shade, tinted blue on its inner surface. The blue rays reflected from it will produce a tolerably pure and white light, by their union with the reddish-yellow rays of the flame. To effect the same purpose the lamp should be enveloped by a glass chimney, tinged inside with a very pale blue, or the light should be made to pass through a fluid of the same color.†

In an early part of this work I have referred in general terms to the important results obtained by Fraunhofer, Kirchhoff, and Bunsen, with reference to the existence of metallic lines ("Fraunhofer's

---

much light, or producing an unpleasant obscurity." "On some of the Forms of Disease of the Eye," by Ernest Hart, Ophthalmic Surgeon to St. Mary's Hospital, etc., p. 21. 1864.

* It has been established by direct experiment that the *violet* ray of the solar spectrum is actually capable of rendering a needle magnetic which has never been touched by the loadstone or by an artificial magnet. (Steinmetz.)

† "Practical Remarks on Near and Aged Sight," by W. White Cooper, Esq., F.R.C.S., England. 1847.

lines ") observed in the sunlight by means of the spectroscope.* It has naturally been a matter of discussion with medical men engaged in sanitary investigations, whether the "vapor of iron," found to exist in conjunction with other metals in the sun's atmosphere, has any hygienic effect upon the physical organism as the result of the therapeutic action of iron on the blood distributed so freely through the minute capillary vessels ramifying on the surface of the body? In the absence of any hypothesis of a more satisfactory character to account for the beneficial action of light, it is rea-

---

* "We can no longer consider light as merely consisting of infinitesimal *particles*, or as infinitesimal *waves;* we may now conclude that it is *metallic;* that sunshine consists of a metallic 'shower,' such as the Greek mythology ascribed to Jupiter in one of his unions with mortals (strange coincidence of fact with emblematic fiction!) but instead of a shower of gold, according to the figment, the beneficent sunshine bathes us with elementary iron, sodium, magnesium, calcium, chromium, nickel, barium, copper, zinc, and hydrogen! It is Jupiter, not Apollo, that the ancient mythology should have identified with the sun. In the idea of *Zeus* (or 'The Burner') and the *Zeuspater* (Father Jove) of the Latins, we might translate all the amours of the Father of the Gods into physical facts connected with the 'arch-chemic sun,' as Dante calls him, and recognize the idea that all animals are children of the sun."—"Sunshine and Showers," by Andrew Steinmetz, Esq., of the Middle Temple, Barrister-at-Law. London, 1867.

sonable to suppose that the iron vapor detected in the sun's beams may have a physiological as well as a mechanical effect upon the composition of the blood, by throwing into the general circulation, through the vessels of the skin, a most important vital constituent.*

In the present state of our knowledge it is impossible to dogmatize upon this subject. The matter in review must still be considered *sub judice*, and purely speculative, and until further experiments are made it is wise to hold in reserve any theory that may have been formed in relation to it.

---

* There does not appear to be anything improbable in this hypothesis, when we consider that mercury may be made to affect the system by subjecting the patient to its influence in the form of a vapor-bath, the mercury of course being absorbed principally by the skin, although a portion may pass into the lungs. That persons handling the soluble salts of lead are often poisoned in consequence of the metal passing through the integuments, is a generally admitted fact.

Dr. Th. Clemens, of Frankfort-on-the-Main, (Deutsche Klinik, 1865-6, Schmidt's Jahrbücher, 1866,) has " observed the immunity from cholera of coppersmiths. Hence he recommends as a cholera disinfectant a spirit of chlorate of copper, and advises the same preparation to be used both internally, and externally upon the skin as an actual preservative."—*Half-Yearly Abstract*, 1866, p. 5.

APPENDIX.

I. THE BLESSINGS OF LIGHT: MILTON'S BLINDNESS.

II. SUGGESTIONS FOR THE REGULATION OF LIGHT IN DARK ROOMS, ETC.

III. THEORIES OF LIGHT.

IV. EFFECT OF LIGHT ON THE SKIN.

V. THE SUN.

VI. NUTRITION OF PLANTS.

VII. VITALITY OF SEEDS.

VIII. THE MOON.

IX. ANIMALS INFLUENCED BY THE STATE OF THE WEATHER.

X. ON THE DISTRIBUTION OF LIGHT BY ARTIFICIAL MEANS.

XI. ADVANTAGES OF LIGHT IN THE TREATMENT OF THE SICK.

XII. ON THE REGULATION OF THE QUANTITY OF LIGHT ADMITTED INTO THE CHAMBERS OF THE SICK.

THE BLESSINGS OF LIGHT: MILTON'S BLINDNESS.

(Vide Preface.)

In the subjoined lines Milton alludes with touching pathos to his own sad deprivation of sight:

> " Thus with the year
> Seasons return, but not to me returns
> Day, or the sweet approach of eve or morn,
> Or sight of vernal bloom, or summer's rose,
> Or flocks, or herds, or human face divine;
> But cloud instead, and ever-during dark
> Surrounds me, from the cheerful ways of men
> Cut off, and for the brook of knowledge fair
> Presented with an universal blank
> Of Nature's works, to me expunged and rased,
> And knowledge at one entrance quite shut out."

I append to the preceding affecting wail of distress a copy of a posthumous poem (having reference to his blindness) *said* to have been written by Milton. I have not been able to ascertain at what period these lines were penned; but assuming them to be genuine, I should suppose they were written long subsequently to the verses previously quoted.* They are deeply interesting, as showing the

---

[* The American Publishers deem it proper to state that this poem was not written by Milton, but by Elizabeth Lloyd, a Quakeress of Philadelphia. In reprinting it here, they have deemed it proper to follow the version published in the *National Magazine*, vol. viii, 1856, p. 85, which is more correct than that of Dr. Winslow. Elizabeth Lloyd died in the latter part of the 18th century.]

state of Milton's feelings with regard to his infirmity at different epochs of his life :

I.

"I am old and blind!
Men point to me as smitten by God's frown;
Afflicted and deserted of my kind—
 Yet I am not cast down.

II.

"I am weak, yet strong;
I murmur not that I no longer see—
Poor, old, and helpless, I the more belong,
 FATHER SUPREME, to THEE!

III.

"O merciful One!
When men are farthest, then thou art most near;
When friends pass by, my weakness shun,
 THY chariot I hear.

V.

"THY glorious Face
Is leaning toward me, and its holy light
Shines in upon my lonely dwelling place,
 And there is no more night.

V.

"On my bended knee
I recognise Thy purpose clearly shown:
My vision Thou hast dimmed that I may see
 THYSELF—THYSELF alone.

VI.

"I have naught to fear;
This darkness is the shadow of THY wing;
Beneath it I am almost sacred—here
    Can come no evil thing.

VII.

"O! I seem to stand
Trembling, where foot of mortal ne'er hath been,
Wrapped in the radiance of that sinless land,
    Which eye hath never seen.

VIII.

"Visions come and go;
Shapes of resplendent beauty round me throng;
From angel lips I seem to hear the flow
    Of soft and holy song.

IX.

"It is nothing now:
When heaven is opening on my sightless eyes,
When airs from Paradise refresh my brow,—
    That earth in darkness lies.

X.

"In a purer clime
My being fills with rapture; waves of thought
Roll in upon my spirits; trains sublime
    Break over me unsought.

XI.

"Give me now my lyre!
I feel the stirrings of a gift divine;
Within my bosom glows unearthly fire,
    Lit by no skill of mine."

SUGGESTIONS FOR THE REGULATION OF LIGHT IN DARK ROOMS.

(P. 5.)

" In all great towns, where neither houses nor palaces can be isolated, there are almost in every edifice dark and gloomy crypts thirsting for light; and in the city of London there are places of business where the light of day never enters, and where the precious light which the sky sends down between chimney-tops is allowed to fall useless on the ground. On visiting a friend whose duty confined him to his desk during the official part of the day, we found him with bleared eyes struggling against the feeble light which the opposite wall threw into his window. We counselled him to extend a blind of fine white muslin on the outside of his window, and flush with the wall. The experiment was soon made. The light of the sky above was caught by the fibres of the linen, and thrown straight upon his writing-table, as if it had been reflected from an equal surface of ground glass. We may mention another case equally illustrative of our process. A party visiting the mausoleum of a Scottish nobleman wished to see the gilded receptacles of the dead which occupied its interior. There was only one small window through which the light entered, but it did not fall upon the objects to be examined. Upon stretching a muslin handkerchief from its four corners, it threw such a quantity of light into the crypt as to display fully its contents." SIR D. BREWSTER.

THEORIES OF LIGHT.

(P. 9.)

I have referred to the two theories that have prevailed in the scientific world regarding the nature and propagation of light, viz., the corpuscular theory of the immortal Newton, and the undulatory theory of Huyghens, afterwards revived and fully elaborated by Dr. Young. Sir Isaac Newton supposed that light was composed of particles of excessive minuteness, which were projected from the luminous body with a velocity equal to 193,000 miles in a second. The undulatory theory was based upon the hypothesis, that light was transmitted from its source by vibrations or undulations of an etherial fluid of great elasticity, in spherical superficies in the form of waves, similar to the transmission of sound.

The principles of the undulatory theory are thus stated by Sir J. Herschel:

1. That an excessively rare, subtle, and elastic medium, or *ether*, fills all space, and pervades all material bodies, occupying the intervals between their molecules; and either by passing freely among them, or by its extreme rarity, offering no resistance to the motion of the earth, the planets, or comets in their orbits appreciable by the most delicate astronomical observations, and having inertia, but not gravity.

2. That the molecules of the ether are susceptible of

being set in motion by the agitation of the particles of ponderable matter; and that when any one is thus set in motion it communicates a similar motion to those adjacent to it, and thus the motion is propagated farther and farther in all directions, according to the same mechanical laws which regulate the propagation of undulations in other elastic media, as air, water, or solids, according to their respective constitutions.

3. That in the interior of refracting media the ether exists in a state of less elasticity, compared with its density, than *in vacuo* (*i.e.*, in space empty of all other matter); and that the more refractive the medium, the less, relatively speaking, is the elasticity of the ether in its interior.

4. That vibrations communicated to the ether in free space are propagated through refractive media by means of the ether in their interior, but with a velocity corresponding to its inferior degree of elasticity.

5. That when regular vibratory motions of a proper kind are propagated through the ether, and, passing through our eyes, reach and agitate the nerves of our retina, they produce in us the sensation of light in a manner bearing a more or less close analogy to that in which the vibrations of the air affect our auditory nerves with that of sound.

6. That as, in the doctrine of sound, the frequency of the aerial pulses, or the number of excursions to and fro, from the point of rest made by each molecule of the air, determines the pitch, or note; so, in the theory of light, the frequency of the pulses, or number of impulses made on our nerves in a given time by the etherial molecules next

in contact with them, determines the *color* of the light; and that as the absolute extent of the motion to and fro of the particles of air determines the *loudness* of the sound, so the *amplitude* or extent of the excursions of the etherial molecules from their points of rest determines the brightness or intensity of light.—*Dictionary of Science, Literature, and Art.* By W. T. Brande, D.C.L., F.R.S.L. & E.

EFFECT OF LIGHT ON THE SKIN.

(P. 14.)

"Amongst the carpenters I noticed one with a remarkable staining of the skin, some portions of his face and neck being intensely bronzed, while other parts remained quite fair, giving the man a very unsightly mottled appearance. In plain terms, he was enormously freckled, the stained skin being about the average color of the faces of the artisan and laboring classes in the north of China. His disfigurement, consequently, arose from the refusal of the whole of the exposed portions of his body equally to take on the bronzing process. I made him show me his skin where habitually protected from the sun, and it was as fair as a European's."—*Peking and the Pekingese.* By D. F. Rennie, M.D. London. 1865.

THE SUN.

(P. 45.)

"The sun
Had first his precept so to move, so shine,
As might affect the earth with cold and heat
Scarce tolerable, and from the north to call
Decrepit winter, from the south to bring
Solstitial summer's heat. To the blank moon
Her office they prescribed; to the winds they set
Their corners, when with bluster to confound
Sea, air, and shore; the thunder when to roll
With terror through the dark aerial ball. . . .
Beast now with beast 'gan war, and fowl with fowl,
And fish with fish: to graze the herb all leaving,
Devoured each other, nor stood much in awe
Of man, but fled him, or with countenance grim
Glared on him passing."—MILTON.

"The sun's rays are the ultimate source of almost every motion which takes place on the surface of the earth. By its heat are produced all winds, and those disturbances in the electric equilibrium of the atmosphere which give rise to the phenomena of lightning; and probably, also, to those of terrestrial magnetism and the aurora. By their vivifying action vegetables are enabled to draw support from inorganic matter, and become, in their turn, the support of animals and of man, and the sources of those great deposits of dynamical efficiency which are laid up for human use in our coal strata. By them the waters of the sea are made to circulate in vapor through the air, and irrigate the land, producing springs and rivers. By them are produced all disturbances of the chemical equilibrium of the elements of nature, which, by a series of compositions and decompositions, give rise to new products, and originate a transfer of materials. Even the slow degradation of the solid constituents of the surface, in which its chief geological changes consist, is almost entirely due, on the one hand, to the abrasion of wind and rain, and the alternation of heat and frost; on the other, to the continual beating of the sea waves, agitated by winds, the result of solar radiation. Tidal action (itself partly due to the sun's agency) exercises here a comparatively slight influence. The effect of oceanic currents (mainly originating in that influence), though slight in abrasion, is powerful in diffusing and transporting the matter abraded; and when we consider the immense transfer of matter so produced, the increase of pressure over large spaces in the bed of the ocean, and diminution over corresponding portions of the land, we are not at a loss to perceive how the elastic power of subterranean fires, thus repressed on the one hand, and relieved on the other, may break forth in points where the resistance is barely adequate to their retention, and thus bring the phenomena of even volcanic activity under the general law of solar influence.

"The great mystery, however, is to conceive how so enormous a conflagration (if such it be) can be kept up. Every discovery in chemical science here leaves us completely at a loss, or rather, seems to remove farther the prospect of probable explanation. If conjecture might be hazarded, we should look rather to the known possibility of an indefinite generation of heat by friction, or to its excitement by the electric discharge, than to any actual combustion of ponderable fuel, whether solid or gaseous, for the origin of solar radiation."—*Good Words.*

NUTRITION OF PLANTS.
(P. 53.)

"Since the sun-light is composed of many differently colored rays and different principles, it becomes an interesting inquiry which of these is the immediate agent in ministering to the nutrition of plants. In 1843, by causing plants to effect the decomposition of carbonic acid in the prismatic spectrum, I found that the yellow is by far the most effective, the relative powers of the various colors being as follows: 1, yellow; 2, green; 3, orange; 4, red; 5, blue; 6, indigo; 7, violet."—DRAPER.

VITALITY OF SEEDS.
(P. 54.)

It has been a question with botanists how to account for the longevity or vitality of seeds: I refer particularly to wheat, which has been found in Egyptian sarcophagi, proved to be many thousand years old. These seeds, when put in the ground, have fully germinated, and been most productive. The persistent vitality of such seeds, according to the best botanical authorities, depends upon three principal conditions, viz.: 1, uniformity of temperature; 2, moderate dryness; and 3, the exclusion of light. Seeds brought from India (says Dr. Lindley) round by the Cape of Good Hope rarely vegetate freely. This he says is owing to the double exposure to the heat of the equator, and the subsequent arrival of the seeds in cold latitudes. Seeds brought overland from India, and not exposed to such fluctuations of temperature, retain their active vital principle. Seeds will travel with safety for many months if buried in clay rammed hard in boxes. Seeds of the mango are thus brought alive from the West Indies, as well as the principal part of the araucania pines which have been transported from Chili to England.

THE MOON.

(P. 59.)

The moon appears to have called forth the fire and sublimity of poetic genius in all ages and in all climes. Some of the most beautiful and touching sonnets that adorn the English language are addressed to the moon. I cannot forbear to quote an illustration from the pen of Charlotte Smith, one of our most exquisite writers of sonnets, the immortal Milton alone excepted:

> "Queen of the silver bow! by thy pale beam,
> Alone and pensive, I delight to stray,
> And watch thy shadow trembling in the stream,
> Or mark the floating clouds that cross thy way;
> And while I gaze, thy mild and placid light
> Sheds a soft calm upon my troubled breast:
> And oft I think—fair planet of the night—
> That in thy orb the wretched may have rest;
> The sufferers of the earth, perhaps, may go,
> Released by death, to thy benignant sphere;
> And the sad children of despair and woe
> Forget in thee their cup of sorrow here.
> Oh, that I soon may reach thy world serene!
> Poor wearied pilgrim in this toiling scene!"

"In savage life, especially in high latitudes, the moon is an ever-present power. When the Red Indian speaks of moons as measures of time, he speaks in the tone of affection and reverence for the benign luminary that guides his steps through the trackless forest. The Oriental bows to the sun, but the Red Indian nurtures his grand and impassive nature in the mild beams of the moon. In hunting and trapping, the moon is his ever-faithful ally, and he would as soon think of doubting its use as he would the use of his spear or his traps.

"But the use of the moon is not confined to light-giving. As a mechanical power, the moon is of much service. The sun is the grand source of power on the face of the earth; but still some little work is left for the moon. To her chiefly is assigned the task of raising the tides of the ocean. The tides are of incalculable benefit to man. In a sanitary point of view, the moon may be regarded as the great scavenger of our globe. Twice every day, she flushes, with sea-water in abundance, the rivers on which our towns are situated, and keeps them comparatively pure. Again, by her mechanical power, she bears ships on the crest of the tidal wave deep into the heart of the country, where the centres of commerce are often found. Insignificant streams are thus rendered navigable, and cities brought into immediate connection with the ocean—the highway of commerce. By the convenience afforded by the moon, London is at the same time connected with the ocean, and in the heart of the country, where it can be best protected from any invasion. In an island of such limited extent as Great Britain, the rivers must necessarily be small, but the tidal wave compensates for the defect, and gives us the advantages of river navigation. The mechanical power of the tide is made available by means of the tide-mill. The rise and fall of the tide can be utilized as well as the fall of the river. This source of power has not been very generally turned to account, though there is no mechanical difficulty in applying it."—*Good Words*.

ANIMALS INFLUENCED BY THE STATE OF THE WEATHER.

(P. 149.)

MANY of the diseases of animals are greatly influenced by states of the atmosphere. It is remarked that immediately before rain, and particularly before great falls of snow in winter, dogs are dull: their ears become inflamed, and they lie drowsily by the fire the principal part of the day. Swine are observed to be uneasy in windy weather, and show symptoms of restlessness even before the winds begin to blow by running about with a peculiar tossing up of the head. Hence the popular notion that pigs can foresee the wind.

ON THE DISTRIBUTION OF LIGHT BY ARTIFICIAL MEANS.

(P. 156.)

With the object of distributing light, Sir David Brewster suggests that "the opposite sides of the street or lane should be kept whitewashed with lime, and for the same reason the ceilings and walls of the apartments should be as white as possible, and all the furniture of the lightest colors. Having seen such effects produced by such imperfect means, we feel as if we had introduced our poor workman or needle-woman from a dungeon into a summer-house, where the aged can read their Bible, where the inmates can see each other, and carry on their work in facility and comfort.

ADVANTAGES OF LIGHT IN THE TREATMENT OF THE SICK.

(P. 167.)

" Who has not observed the purifying effect of light, and especially of direct sunlight, upon the air of a room?

Here is an observation within everybody's experience. Go into a room where the shutters are always shut, (in a sick-room or a bed-room there should never be shutters shut,) and though the room be uninhabited, though the room has never been polluted by the breathing of human beings, you will observe a close, musty smell of corrupt air, *i. e.*, unpurified by the effect of the sun's rays. The mustiness of dark rooms and corners, indeed, is proverbial. The cheerfulness of a room, the usefulness of light in treating disease, is all-important.

" Heavy, thick, dark window or bed curtains should, however, hardly ever be used for any kind of sick in this country. A light white curtain at the head of the bed is, in general, all that is necessary, and a green blind to the window, to be drawn down only when necessary.

"' Where there is sun there is thought.' All physiology goes to confirm this. Where is the shady side of deep valleys, there is cretinism. Where are cellars and the unsunned sides of narrow streets, there is the degeneracy and weakliness of the human race, mind and body equally degenerating. Put the pale withering plant and human being into the sun, and if not too far gone, each will recover health and spirit.

" It is a curious thing to observe how almost all patients lie with their faces turned to the light, exactly as plants always make their way toward the light. A patient will even complain that it gives him pain 'lying on that side.' 'Then why *do* you lie on that side?' He does not know; but we do. It is because it is the side toward the window.

A fashionable physician has recently published in a Government report that he always turns his patients' faces from the light. Yes, but nature is stronger than fashionable physicians, and depend upon it she turns the faces back, and *toward* such light as she can get. Walk through the wards of a hospital, remember the bedsides of private patients you have seen, and count how many sick you ever saw lying with their faces toward the wall."—*Notes on Nursing*. By Florence Nightingale.

ON THE REGULATION OF THE QUANTITY OF LIGHT ADMITTED INTO THE CHAMBER OF THE SICK.

(P. 169.)

" The quantity of light admitted into the sick-chamber is a matter of immense importance to its suffering occupant. As light is an element of cheerfulness, it is on that account desirable that as much should be admitted as the patient can bear without inconvenience. The light should be soft and subdued, not glaring; and care should be taken that bright, lustrous objects, such as crystals and looking-glasses, should be kept out of the patient's view, and that neither the flame of a lamp or candle, nor its reflection in a mirror, be suffered to annoy him by flashing across his field of vision."—RIDGE *on Health and Disease.*

THE END.

THE
ATTENTION OF THE MEDICAL PROFESSION
IS CALLED TO THE FOLLOWING
MEDICAL JOURNALS
PUBLISHED BY US.

THE NEW YORK MEDICAL JOURNAL. Edited by WILLIAM A. HAMMOND, M.D., Professor of Diseases of the Mind and Nervous System in the Bellevue Hospital Medical College, and E. S. DUNSTER, M.D., Physician to the Out-door Department of the Bellevue Hospital. Issued monthly. Terms, $5.00 per annum.

THE QUARTERLY JOURNAL OF PSYCHOLOGICAL MEDICINE AND MEDICAL JURISPRUDENCE. Edited by Professor WILLIAM A. HAMMOND, M.D. Issued quarterly. Terms, $5.00 per annum.

THE MEDICAL GAZETTE. A WEEKLY REVIEW of Practical Medicine, Surgery and Obstetrics. Edited by LEROY MILTON YALE, M.D., Physician to the Out-door Department of Bellevue Hospital. Issued weekly. Terms, $2.00 per annum.

The sixth volume of the NEW YORK MEDICAL JOURNAL began with the October number, 1867.

The first volume of the PSYCHOLOGICAL JOURNAL began with the July number of that year. Its success is so great that the number has been reprinted.

The first number of the MEDICAL GAZETTE was issued Sept. 28th, 1867. It consists of 8 large two-column pages, printed in excellent style.

When two or more of these Journals are subscribed for, they will be furnished at the following rates:

THE NEW YORK MEDICAL JOURNAL and the PSYCHOLOGICAL JOURNAL..$8 00
Either of the above and the GAZETTE............................ 6 00
All three of the Journals 10 00

Any new subscriber to either the NEW YORK MEDICAL JOURNAL or the PSYCHOLOGICAL, remitting us $5.00 for a year's subscription, will receive, post-paid, a copy of *Murray on Emotional Diseases of the Sympathetic System of Nerves*, the price of which alone is $1.50.

Payment in all cases must be *in advance*.

MOORHEAD, SIMPSON & BOND,
PUBLISHERS,
No. 60 Duane Street, New York.

ON CHRONIC ALCOHOLIC INTOXICATION. With an inquiry into the influence of the abuse of Alcohol, as a predisposing Cause of Disease. By W. MARCET, M.D., F.R.S., Fellow of the Royal College of Physicians, etc., etc. In convenient size, large type. Will be issued in a few days, by

MOORHEAD, SIMPSON & BOND,

60 DUANE STREET,

NEW YORK.

THE DARTROUS DIATHESIS, OR ECZEMA AND ITS ALLIED AFFECTIONS. By A. HARDY, M.D., physician to the Hospital St. Louis, Paris. Translated by HENRY G. PIFFARD, M.D., late House Surgeon Bellevue Hospital; Physician to the Skin Department, Out-Door Bureau, Bellevue Hospital; Assistant to Chair of Theory and Practice, Bellevue Hospital Medical College.

All orders should be addressed,

MOORHEAD, SIMPSON & BOND,

PUBLISHERS AND PRINTERS,

60 Duane Street, New York.

HYSTERIA.—REMOTE CAUSES OF DISEASE IN GENERAL. TREATMENT OF DISEASE BY TONIC AGENCY. LOCAL OR SURGICAL FORMS OF HYSTERIA, etc. Six Lectures delivered to the Students of St. Bartholomew's Hospital, 1866. By F. C. SKEY, F.R.S.

The well-known ability and experience of the distinguished author of this work will insure for it the respectful consideration of the Medical Profession. It is truly admirable, both in form and substance, and will doubtless accomplish much good through those who follow its progressive and scientific teachings.

Sent free by mail on receipt of price, $1.50.

MOORHEAD, SIMPSON & BOND, Publishers,

No. 60 DUANE STREET, NEW YORK.

A TREATISE ON

EMOTIONAL DISORDERS OF THE SYMPATHETIC SYSTEM OF NERVES. By WM. MURRAY, M.D., etc., Physician to the Dispensary, and to the Hospital for Sick Children, and Lecturer on Physiology in the College of Medicine, Newcastle on Tyne.

CONTENTS.

THE VARIETIES OF EMOTION—Visceral Sensations.
THE EFFECTS OF EMOTION ON THE CEREBRO-SPINAL SYSTEM.
THE EFFECTS OF EMOTION ON THE SYMPATHETIC SYSTEM—The Heart. The Arteries. Involuntary Muscles. Secreting Glands. The Stomach. Dyspepsia. The Liver. The Urinary Organs. The Generative and Sexual Organs. Hysteria.
PREDISPOSING CAUSES OF EMOTIONAL DISORDERS.
DISEASES OF THE DIGESTIVE ORGANS—Cases. The Stomach. Visceral Sensations and Gasteria. The Liver and Hypochondriasis.
ON THE MODUS OPERANDI OF DYSPEPSIA — Explanatory Theory. Imaginary Diseases.
THE ORGANS OF GENERATION—Nymphomania. Masturbation. Hysteria.
THE CHANGE OF LIFE—Symptoms connected with the Sympathetic System. Flushes, &c. Emotional Symptoms. Uterine Affections. Miscarriages. Disordered Action of the Colon from Ganglionic Sympathy with the Uterus.
THE MALE ORGANS.
TREATMENT—To prevent Emotion from injuring the Body. To relieve Injuries produced by severe Emotion. Remarks on the General Management of Cases. Stimulants: Camphor, Naphtha, Aromatics, Anti-spasmodics, The Fœtid Gums, Valerian, Sulphur. Tonics: Iron, Nux Vomica, Quinine, the Salts of Silver, Zinc. Local Applications: The Action of Cold Water Compresses. Summary.

Extra Cloth. Price **$1.50**. Sent by mail on receipt of price.

MOORHEAD, SIMPSON & BOND,

No. 60 DUANE STREET, NEW YORK.

SLAVE SONGS OF THE UNITED STATES. Set to Music. A new edition of this popular work is just issued. Price $1.50.

The Press says of it:

"We welcome the volume before us—the first collection of negro songs, words and music, that has been made."—*N. Y. Independent.*

"They have no sort of resemblance to the so-called negro songs of the cork minstrels, and, as a rule, are much more attractive."—*N. Y. Citizen.*

"The verses are expressive, and the melodies touching and effective. There is no doubt the book will receive wide attention."—*Brooklyn Standard.*

"Possesses a curious interest for the student of African character."—*N. Y. Tribune.*

"These endemical lays are, in fact, chief among the signs and evidences of the normal African character."—*N. Y. World.*

"This collection contains many excellent ballads that might readily be supposed the work of our best composers."—*Le Messager Franco-Americain.*

MOORHEAD, SIMPSON & BOND,
PUBLISHERS,
No. 60 DUANE STREET, NEW YORK.

LIGHT: ITS INFLUENCE ON LIFE AND HEALTH. By FORBES WINSLOW, M.D., etc.

This important work is written in the most simple and perspicuous manner, and is one of those contributions to science which all can understand and appreciate. It is, indeed, a charming treatise on a subject of vital interest.

NEARLY READY.

GALILEO: HIS LIFE, HIS DISCOVERIES, AND HIS WORKS. Translated from the French of Dr. Max. Parchappe, by HUON D'ARAMIS.

This biography of one of the most celebrated men the world has produced, has been pronounced, by European critics, to be the best ever written. Dr. Max. Parchappe, the author, died just as he had completed a work which will long remain a monumen of his erudition and impartiality.

Notice to Medical Authors.

Messrs. Moorhead, Simpson & Bond, Proprietors of the Agathynian Press, respectfully call the attention of Medical Authors to their facilities for publishing, in handsome style, and bringing before the Profession through their several journals, Medical Works, such as Monographs, Text-Books, Essays, Lectures, &c., &c.

The extensive business connections of Messrs. Moorhead, Simpson & Bond, and their reputation for publishing good works, enable them to command a ready sale for all their publications.

The typographical and literary excellence of the journals and works issued by them, have attracted the marked attention of the press and of the reading public.

Estimates promptly furnished upon application.

They are the printers and publishers for the Agathynian Club, an association of gentlemen devoted to elegant typography.

MOORHEAD, SIMPSON & BOND,

60 Duane Street,

NEW YORK.

THE PRINCIPLES AND PRACTICE OF LARYNGOSCOPY AND RHINOSCOPY in Diseases of the Throat and Nasal Passages. Designed for the use of Physicians and Students. With 59 Engravings on Wood. By ANTOINE RUPPANER, M.D., M.A., Member of the American Medical Association; of the Massachussetts Medical Society; of the County Medical Society of New York, etc. Sent by mail, postage paid, on receipt of price. $2.00.

MOORHEAD, SIMPSON & BOND,

PUBLISHERS AND PRINTERS,

60 Duane Street, N. Y.

A BOOK THAT NO PHYSICIAN SHOULD BE WITHOUT.

PATHOLOGICAL ANATOMY OF THE FEMALE SEXUAL ORGANS, By Prof. JULIUS KLOB, of Vienna. Translated by Drs. J. KAMMERER and B. F. DAWSON. First volume, containing the Diseases of the Uterus. Extra cloth, Price, $3.50.

MOORHEAD, SIMPSON & BOND,

PUBLISHERS,

60 Duane Street, N. Y.

www.ingramcontent.com/pod-product-compliance
Lightning Source LLC
Chambersburg PA
CBHW021304240426
43669CB00042B/1225